# Genetic Essentialism

This book examines how science education can either reinforce or counteract genetic essentialism, a misperception that social groups share genes which determine their cognitive and behavioral characteristics.

The first chapter examines genetic essentialism's key features, current manifestations and impacts, and historical precedents. The middle chapters analyze how science textbooks and curricula enable and may amplify essentialist reasoning through simplified explanations of genetics, reproductive biology, and human evolution. These chapters build a more accurate view through research in fields including anthropology, biology, education, history of science, psychology, and sociology. The final chapter compares egalitarianism and stratification across evolutionary time scales to demonstrate that inequality and essentialism are interrelated and socially constructed. The overall aim is to spotlight genetic essentialism in instructional resources and refute it through evidence and education.

The book is intended for scholars and students in the human sciences, biological sciences, health and medicine, and journalism, as well as readers in any field who are interested in confronting the conceptual underpinnings of racism and sexism.

**Elizabeth D. Whitaker** is a biocultural anthropologist specializing in human health, social history, and the history of scientific ideas. She lectures at the University of Bologna and has also taught at several universities in the United States. Among her academic honors are three Fulbright awards for research and teaching in Italy. Her publications include four other Routledge titles: *The trouble with human nature: Health, conflict, and difference in biocultural perspective; Health and healing in comparative perspective; A Macat analysis of Marcel Mauss's The gift: The form and reason for exchange in archaic societies;* and, *A Macat analysis of Betty Friedan's The feminine mystique.*

# Genetic Essentialism

## How Science Education Shapes Perceptions of Race and Gender

Elizabeth D. Whitaker

Routledge
Taylor & Francis Group
NEW YORK AND LONDON

First published 2027
by Routledge
605 Third Avenue, New York, NY 10158

and by Routledge
4 Park Square, Milton Park, Abingdon, Oxon, OX14 4RN

*Routledge is an imprint of the Taylor & Francis Group, an informa business*



ISBN: 978-1-041-07316-1 (hbk)
ISBN: 978-1-041-07318-5 (pbk)
ISBN: 978-1-003-63992-3 (ebk)

DOI: 10.4324/9781003639923

Typeset in Sabon
by SPi Technologies India Pvt Ltd (Straive)

# Contents

# 1 Genetic essentialism

## Definitions, impacts, history

Stratified societies need a theory for the inequalities between social groups. Essentialism provides a beguiling, seemingly obvious explanation: an invisible, inborn cause that permanently and predictably determines a group's physical and mental characteristics. The unseen essence may be cosmic, cultural, or biological. Biological essentialism rings through political rhetoric describing immigrants as "poisoning the blood" of a nation, binary sex as the determinant of gender, and the populations of disvalued nations as dull and depraved due to "bad genes." The underlying assumption of shared biological essences ultimately rooted in genes is both scientifically inaccurate and socially injurious. Genetic essentialism misrepresents the role of genes in the development of traits and exaggerates within-group uniformity and between-group divergence. By equating social dominance with innate superiority, genetic essentialism renders inequalities in education, society, and the workplace natural and morally acceptable.

As shown in the following chapters, essentialist biases make it more difficult for students to learn about evolutionary processes and biological variation. Genetic essentialism promotes intergroup hostility, in-group bias, approval of segregation and eugenic policies, and disapproval of or disinterest in personal and public efforts to address disparities. In scientific and medical research and practice, it leads to imprecise analyses and inappropriate diagnosis and treatment, while the determinism inherent in genetic essentialism fuels unrealistic expectations for magic-bullet solutions to complex health and behavioral issues. Misappropriated research findings on health disparities and achievement gaps fuel essentialist claims about inborn race and sex differences as the cause of social and economic inequalities, including imbalances in educational and professional achievement. Innatism regarding scientific ability justifies disparities in prestigious, economically remunerative scientific fields that hinder the advancement of knowledge. These combined impacts affect everyone, for unequal societies generate worse health and reduced wellbeing even among those at the top of the hierarchy, compared to more equal societies.[1]

DOI: 10.4324/9781003639923-1

Science education inadvertently advances genetic essentialism through simplified explanations of genetics, reproduction, and human evolution that reinforce the misconceptions students encounter outside the classroom. Agreement with genetic essentialism rises with increasing grade level, indicating both that it is learned and that it can be unlearned or averted. To help build a more accurate understanding of human variation and thereby refute the genetic essentialism at the heart of racism and sexism, this book combines an analysis of science education with a response drawn from current research in multiple disciplines.

While recognizing that essentialism also contributes positively to identity and in-group cohesion, the following chapters focus on the way genetic essentialism favors exclusion and discrimination by distorting perception of human variation, obscuring the biological impacts of racism and sexism, and interfering with learning. The discussion centers on race/ethnicity and sex/gender, even though essentialist thinking may be leveraged against other groups, such as those defined by disability, religious affiliation, nationality, or political persuasion. The reason is that race/ethnicity and sex/gender are topics that appear consistently in science textbooks in a manner that does not oppose, and may even magnify, essentialist biases.

The current chapter explores the circulation of genetic essentialism through informal channels, whereas the remainder of the book focuses on formal instruction in biology and allied fields. The first three chapters are weighted toward race essentialism, whereas the second three concentrate on sex essentialism, although the themes overlap throughout. The final chapter examines egalitarian versus hierarchical behavior and sociopolitical organization to call into question the naturalness of inequality and associated ideologies.

## Sources and drivers of essentialist beliefs

Perceiving the world through categories and root causes is such a common conceptual bias that it may represent a universal form of human cognition (see Dar-Nimrod and Heine 2011; Lo and Sasaki 2024). Categorization is an efficient way to understand and live in the world and is fundamental to science. In biology, taxonomies of living organisms rest upon essentialist assumptions, yet the criteria are explicit, known to be imperfect, and subject to change. At issue in this book is a form of essentialism that is implicit, falsely based on scientific principles, immutable, and socially consequential.

The idea that social groups (national, ethnic, counter-cultural, etc.) share ways of thinking and behaving derived from a permanent essence comes to the surface when people are surprised to have trouble learning the language of forebears from another country, or restaurant customers are disappointed to find their sushi was made by a non-Japanese chef. Many religious traditions including Judaism, Islam, and Christianity construe women and men as essentially unalike. Some posit a divine origin for different races or castes of humankind, such as Christian polygenism (see below) and the Hindu creation myth that roots four castes of people – as well as the sun, moon, wind, and earth – in the body parts of the original divine being, Purusha. Categorical schemes distribute people socially and economically, delimiting their spatial range and field of acceptable relationships. They generate unconscious biases so powerful that judges of an audition for the orchestra believe they hear an inferior performance if they see a musician with Asian facial features play a classical Western piece, or a woman play a "masculine" instrument or martial music – an effect which disappears if the musician performs behind a screen.[2]

Essentialist reasoning incorporates four interrelated assumptions, listed in Figure 1.1: internal essence, determinism, uniformity, and discreteness (Donovan 2017; Donovan et al. 2021; Donovan et al. 2024b). Groups categorized as similar are thought to share a natural, unchanging, invisible essence. The essence directly and invariably determines their physical appearance as well as cognitive and behavioral traits and is passed intact across generations. Internally homogeneous, groups are seen as distinct from others, with minimal or no overlap in traits.

Genetic essentialism holds that a fixed gene-set is the root cause of an individual's characteristics and is readable from them (see Moore et al. 2025). Genetic determinism connects the two: genes are envisioned as directly and permanently causing traits, independently or nearly independently of extra-genetic factors. Genetic reductionism boils phenotypes down to genotypes, justifying a primary focus on genes as opposed to developmental and experiential processes that influence how genes are expressed.

Applied to groups, genetic essentialism transforms social categories into genetic ones, as exemplified by commercial ancestry testing, and would more appropriately be referred to as "sociogenetic essentialism" (Kampourakis and Fux 2025). For ease of reading, this book retains the shorter term, even though in most instances it refers to beliefs about

| Genetic essentialism concepts | | | |
|---|---|---|---|
| internal essence<br>*shared genes within groups* | determinism<br>*genes directly cause traits* | discreteness<br>*social groups bounded and separate* | uniformity<br>*social groups internally homogeneous* |

*Figure 1.1* Genetic essentialism concepts.

groups rather than individuals. In addition, this book does not use separate terms for essentialism concerning physical traits (bio-somatic essentialism) as opposed to mental traits (bio-behavioral or neurogenetic essentialism), given that the first bleeds into the second where differences in reproductive anatomy or skin color are taken to model disparities in intelligence, ability, and behavior.

Both textbooks and popular media encourage the conceptual glide from bio-somatic to bio-behavioral essentialism. In the case of textbooks, this takes the form of examples involving population variability in monogenic diseases, a binary explanation of sex determination that confounds sex with gender, and simplified explanations of phenotypic causation that provide a genetic-determinist mechanism for essential similarities within, and differences between, groups. Genetic determinism and essentialism infuse popular media including films about cloning and genomic science; print and online discussions of "race realism" and the "biological reality" of binary sex/gender; and news reports on behavioral genetics that imply single-gene causes of complex behavioral and cognitive traits such as happiness or political preference (see Dasgupta et al. 2024; Nam and Sawyer 2024; Nelkin and Lindee 2010).[3] Companies sell direct-to-consumer testing for genetic variants related to ancestry, physical traits, and disease risks. Social media ecosystems buzz with discussions regarding connections between genes and health, behavior, and personal identity, and especially disease risks and predictive genetic testing, including testing for untreatable diseases (see Allen et al. 2019; Haran and O'Riordan 2018; Pearce et al. 2024). Online content expressing racism, sexism, homophobia, transphobia, Islamophobia, and antisemitism has exploded over the past decade, alongside increased public tolerance of hostile rhetoric and a rapid rise in hate crimes and incidents.[4]

The belief that genetic explanations extend beyond physical traits to personality characteristics, aptitudes, and social and economic outcomes is widespread and cuts across class, race, gender, and political lines (Schneider et al. 2018; Shostak et al. 2009).[5] Quantifying the prevalence of genetic essentialism is tricky, however, because people tend not to openly agree with racist or sexist statements, and they may believe that genetics determines the mental and behavioral traits of some categories but not others. At the same time, it is easy to strengthen agreement with innatism through experiments and educational materials that present information which directly or indirectly encourages biological determinism and essentialism.

The power of word choice has been shown through an experiment in which researchers presented two groups of participants the same text about disease genetics with different introductory statements (Parrott and Smith 2014).[6] One introduction used the deterministic "blueprint" metaphor. The other substituted blueprint with "instruction" to call attention to gene-environment processes. Compared to the instruction metaphor, reading the blueprint version was associated with stronger belief that genes determine health, weaker belief in the role of personal behavior in health and disease, and more positive attitudes about genetics research relevant to human diseases.[7] Genomics has brought increased attention to gene-gene and gene-environment processes, yet new terms continue to convey deterministic thinking. For instance, "genetic architecture" alludes to polygenicity while also implying direct genetic causation of traits. In epigenetics, the use of deterministic verbs suggests direct causation despite the focus on the effects of environmentally induced biochemical changes on genetic expression (see Waggoner and Uller 2015).

Even very tiny associations between genes and behavior are sufficient to strengthen genetic-determinist beliefs. For instance, after being informed that a "warrior gene" (see Chapter 3) increases the risk of violent behavior by as little as 0.5%, research participants are more likely to absolve an aggression-gene-carrying murderer of moral responsibility than participants not informed of the gene (Heine et al. 2019).[8] The increase in willingness to absolve rises with higher risk percentages, although nowhere near in proportion to the 160-times difference between 0.5% and 80%. A boost in deterministic thinking about groupwide propensities for crime also may be experimentally induced through culture-based essentialism, but only biology-based essentialism brings reduced blame for wrongdoing (see Bailey et al. 2021).

Both mentioning race and leaving it out of information on biology increase genetic essentialism, as shown by an experiment in which university students studying communication spent three sessions watching a documentary film about genetically modified organisms and reading headlines, advertisements, and excerpts from news stories that mentioned genetics (Lynch et al. 2008). None of these sources discussed race. At the end of the study, the students expressed increased agreement that genes explain racial differences.

Reading about racial differences in disease risks has the same effect as reading that race is genetically real: increased essentialism as well as greater in-group bias (Heine et al. 2019). Participants in a survey-based experiment were divided into four groups: the first read a news story on racial variation in a gene associated with increased risk of heart disease (the "Backdoor Vignette"); the second read an article on the social construction of race; the third read nothing; and the fourth read a vignette that described race as a biological concept (Phelan et al. 2013). Readers of the Backdoor Vignette expressed increased belief in race essentialism compared to the social-construction and no-reading groups, along with equivalent belief in essential differences between races to the race-is-biological group. Another study which offered participants a Public Service Announcement about heart disease, genetics, and race with or without the labels "Black" and "White" found that the first resulted in increased race essentialism and discrimination compared to the second, or no announcement at all (Condit et al. 2004).

In addition, whether the subject is disease or behavior, news reports that discuss genetics provoke increased essentialism. This has been shown by a study in which participants were divided into three groups, each given a different media story to read (Morin-Chassé 2014). One story concerned cancer genetics; the other two reported on behavioral genetics research, which involves concepts that misleadingly appear to express the genetic contribution to a trait (see Chapter 2). Whether they read about cancer or behavioral traits, afterward participants saw traits beyond those discussed in the stories as genetically determined, including behaviors and aptitudes.

Media stories on behavioral genetics are especially potent sources of increased essentialism given their focus on the role of genes in mental traits, as shown by a study in which research participants were given one of three versions of a news article on behavioral genetics principles (Morin-Chassé 2020).[9] The "cautious" version explained that behavioral genetics concerns variation within populations, not between, and argued against genetic determinism. The intermediate version added high heritability estimates in support of Nicholas Wade's argument that behavioral genetics statistics speak to genes shared within populations that explain their diverse social institutions. Participants given the last two of these readings expressed increased belief in genetic essentialism.

The cautious version resulted in an increase in genetic essentialism among participants without biology training that was balanced by a decrease among those who had studied biology, indicating that greater genetics/genomics literacy may protect against the triggering effect of genetic essentialism in news stories (Morin-Chassé 2020). Similarly, while reading genetic ancestry test results tends to strengthen belief in essential differences between populations (see Phelan et al. 2014), the effect seems to be driven mainly by the reaction of people unfamiliar with genetics. Roth et al. (2020) found no increase in essentialism among White research participants randomly assigned to receive a genetic ancestry test compared to the group that received no test.[10] However, this overall result had two components: a decline in essentialism among participants well-versed in genetics, and an increase in essentialism among those with little genetics knowledge.

Widespread confusion about genetics and genomics indicates ample conceptual space for determinist and essentialist misconceptions. An analysis of 500 student submissions (20% of the total collected over two years) to the US National DNA Day contest found that 55.6% of the essays contained at least one misconception and 20.2% contained two or more (Mills Shaw et al. 2008). In the second-highest category of errors, "patterns of inheritance," 80% concerned basic Mendelian principles, knowledge of which is prioritized by genetics standards throughout the country (see Chapter 2). While more than 70% of 1,200 respondents to a US national genetics survey correctly answered questions about monozygotic twins versus ordinary siblings, fewer than 50% correctly answered all the other questions (Christensen et al. 2010). Knowledge was weakest about the relationship between single genes and behaviors (24% correct) and the genetic basis of race (25% correct).[11] While confirming low genetics knowledge, a survey answered by 2,050 general public respondents and 2,023 participants in a research study on genetics and autism demonstrates that instruction via a research study can make a positive difference (Little et al. 2022). The first group scored lower than the research study group on all three components: genetics familiarity (5.36 versus 5.79 on a 1–7 scale), knowledge of genetics concepts (56.6% versus 66.1% correct answers), and skills in applying them (44.2% versus 59.5% correct answers).

According to a century-old misconception held by a substantial majority of the general population, the genetic differences between human races are comparable to those separating dog breeds (Donovan et al. 2019). On the one hand, the comparison works because both dogs and humans form a single species. On the other, dog breeds differ in appearance and behavior due to selective breeding, for which there is no human equivalent. The amount of genetic variation between human populations is 6–7 times smaller than between dog breeds: less than 5% versus around 30% (Norton et al. 2019). That is, human groups are far more genetically similar than dog breeds, and there is far more genetic variability within categories than between them (see Chapter 3).

## The utility and appeal of essentialism

The foregoing discussion has shown that genetic essentialism is a widespread and readily amplified way of interpreting human variation.[12] Essentialism renders out-groups unrelated to the self and consequently blocks social interest in and empathy toward out-group members, resulting in reduced cross-group interest and interaction and increased approval of existing structural inequalities (Donovan 2022; Lee et al. 2020; Nam and Sawyer 2024). Belief in biological race correlates with reduced interest both in correcting racial disparities and interacting with people from other groups, irrespective of background race prejudice (Williams and Eberhardt 2008).[13] Sex/gender essentialism is associated with greater support for policies, laws, and social institutions that harden gender boundaries, whether the measures are meant to be protective (such as single-gender classrooms) or punitive (such as forcing transgender individuals to identify as female or male on government forms) (Roberts et al. 2017).[14]

As a form of motivated cognition, genetic essentialism can make sense of both low and high social position through ideas about natural differences in ability and worth. For dominant groups in societies such as the US that are both stratified and purport to uphold ideals of equality and fairness, it reconciles the conflict between their advantaged position and egalitarian values (see Brescoll et al. 2013; Diesendruck 2021; Keller 2005). People from dominant groups respond to the suggestion that science supports genetic essentialism with greater self-worth, stronger endorsement of inequality, and reduced acknowledgment of the unequal treatment of subordinated groups. A study involving participants recruited through a psychology research website illustrates these effects (Morton et al. 2009b). Participants were given a text that described the theory that gender differences are biologically based, followed by a summary of evidence that either supported or refuted the argument. Among both women and men, reading the summary supportive of essentialism brought increased agreement with the legitimacy of gender inequality along with reduced belief in the possibility of social change. Among men, but not women, there was greater agreement that women and men are treated fairly in society; more support for discrimination based on gender; and increased self-esteem.

The threat of social change likewise heightens essentialism and approval of discriminatory practices among dominant groups, as shown by a pair of studies in which university students read an invented news story on women's accomplishments relative to men's, followed by an article on genes, brains, and behavior in mice that openly suggested relevance to humans (Morton et al. 2009b).[15] The article on accomplishments described gender inequalities either as stable or narrowing in the first study, whereas it described areas in which women were now advantaged over men in the second. In both cases, increased sexism over the baseline assessment was associated with stronger agreement with gender essentialism and greater support for biologically based research on gender differences only among men, and only when gender inequalities were described as lessening (as opposed to stable or already changed).

In addition serving dominant groups, genetic essentialism may be embraced by other groups in various ways. For instance, women may explain their subordinate social position in terms of essential gender differences. People from underrepresented gender, sexuality, and ethnic/racial groups may endorse essentialist beliefs to favor identity formation, reduce blame for characteristics over which people have no control (such as sexual preference), and promote in-group cohesion and motivation toward activism (see Ryazanov and Christenfeld 2018).

While dominant groups benefit from essentialist ideas that explain their position in the hierarchy of privilege, they reject essentialization of their own group when presented with the prospect of the race-based exclusion they tolerate or endorse in the case of racial minorities. This has been shown by a pair of studies in which White university students were given an announcement about committee decisions concerning candidates in an arts competition, accompanied by a photograph of a racially ambiguous person (Morton et al. 2009a). The text offered a combination of race-based inclusion or exclusion and White or non-White identity (indigenous Australian in the Australia study or Asian ancestry in the British one). In both studies, opposition was greatest to the exclusion of a candidate for being White and strongest among prejudiced participants. The exclusion of a White candidate caused participants to question the validity of the race concept as applied to their own group. Prejudiced Australians suspended their view of race as natural, whereas prejudiced British participants espoused egalitarian values they otherwise rejected.

National surveys in the US indicate that belief in the genetic determination of group traits, including intelligence, general ability, and athleticism (singly or in combination), is stronger among non-White and lower-socioeconomic-status respondents, those with lower educational levels, and people in older age groups (Outram et al. 2018). The age effect may be overstated, however, due to younger people's greater adherence to an ideal of race colorblindness, which denies unconscious bias and obscures systemic racism and its historical impacts (see Lawless and Chen 2018). Increased essentialism is also associated with conservative political and religious convictions, which interfere with learning about human biological variation and increase the likelihood that exposure to genetics and genomics will strengthen essentialist reasoning (see McCartney 2023; Stern et al. 2023).

The effect of conservative ideology on the misinterpretation of information regarding human genetics is evident from a US national survey whose results were divided between the 12.15% of 1,500 respondents who self-identified alt-right supporters on the initial questionnaire and the rest of the pool (Sawyer and Nam 2024). Participants were randomly distributed between a control group that read about summer homework and four other groups that were given news articles about violence: one with a genetic explanation that mentioned race; one with a genetic explanation that did not mention race; one with a socioenvironmental explanation that mentioned race; and, one with a socioenvironmental explanation that did not mention race. Two weeks later, participants completed a second instrument that assessed agreement with essentialist beliefs. The two genetics texts brought increased essentialism across the board, whereas the socio-ecological readings did not affect the main group but spurred greater essentialist thinking in the alt-right group, both where race was mentioned and where it was not. Alt-right supporters were three times more likely to misread socioecological information as indicating the existence of underlying genetic causes of violent behavior.

A study from northeast Brazil provides further evidence of the relationship between conservative beliefs and genetic essentialism. Responses from first-year, mostly non-science public-university undergraduate students to the "Public Understanding and Attitudes towards Genetics and Genomics" (PUGGS) instrument revealed that older age and greater strength of religious belief were stronger predictors of genetic determinist thinking than low genetics and genomics knowledge (Gericke et al. 2017). The authors suggest that the age effect could be due to older students having higher socioeconomic status relative to younger ones, which, as noted above, induces motivated cognition to defend privileges and advantages. Alternatively, the association could be due to increased conservativism as people start families and enter a more conventional phase of life.

Conservative religious beliefs that include creationism favor genetic essentialism through the idea that God created humans in their present form, including distinct and unvarying sexes and races. Nearly three-quarters of 1,024 polled US adults chose divine creation alone or evolution plus God's guidance, whereas only one-quarter agreed that humans evolved from other species through biological processes alone (Gallup.com 2024). International polls indicate that rejection of exclusively biological evolution is less frequent in Latin America, Japan, and 32 European countries except Turkey than in the US, while evolutionary theory is less well-accepted in the UK than Europe (Pobiner 2016). Rates of agreement with exclusively divine origins for humankind are especially high in predominantly Muslim countries in Africa and the Middle East, as shown by an international survey of teachers and teachers-in-training discussed in Chapter 4. Creationist beliefs are more frequent among people who attend religious services regularly; political conservatives; people who have completed fewer years of education; and older people, except in Latin America – where young people are more likely to reject biological evolution, possibly due to switching from Catholicism to evangelical Protestantism (Pobiner 2016).

A survey of over 8,000 teachers and teachers-in-training in 23 countries as part of the European "BIOHEAD-Citizen" project probed deterministic thinking in relation to two main topics: innate intellectual similarity between people such as twins, and inborn differences as the explanation for gender and racial/ethnic disparities (Castéra and Clément 2014).[16] Across countries, greater biology knowledge and more years of education correlated with weaker innatism. Teachers in Australia, Brazil, and European countries, including France and Finland, expressed less deterministic language and less sexism and racism than those in Lebanon and North African countries (Morocco, Algeria, Tunisia). These results indicate that conservative religious beliefs are sufficient to sustain genetic essentialism independently of exposure to the genetics-saturated mediascape and medical/research sector typical of wealthier countries.

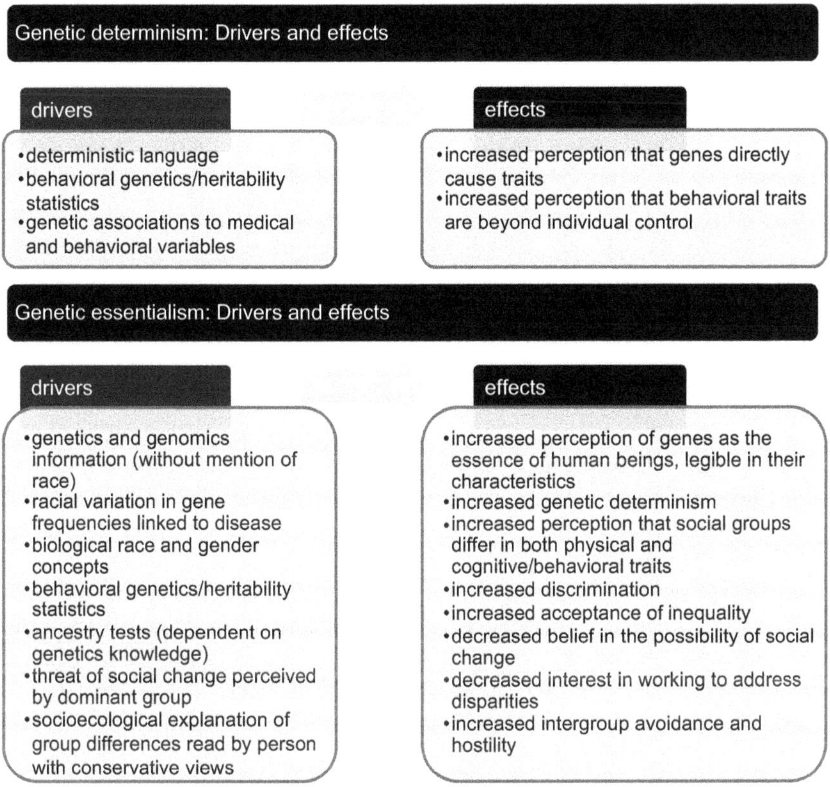


*Figure 1.2* Drivers and effects of genetic determinism and essentialism.

Figure 1.2 summarizes some of the drivers and effects of genetic determinism and essentialism surveyed above.

## Development of essentialist thinking

The tendency to see shared essences in similar items emerges early in life as children attempt to make sense of their observations of the world. By age two, they have a solid sense of personal gender identity, which manifests itself in preferences for culturally coded toys, activities, attire, and behavior. Within a year or two, they become aware of gender stereotypes and express gender biases in line with the variable family and sociocultural contexts in which they are raised (King et al. 2021). By age four or five, children have internalized essentialist thinking about social categories as natural distinctions between sex/gender and race/ethnicity categories and conceptualize gender in animals the same way as humans (see Meyer and

Gelman 2016; Rhodes and Mandalaywala 2017). Consciousness of stereotypes rises quickly between the age of 6 and 10 (McKown and Weinstein 2003), which is the same period during which children come to envision gender differently in humans than animals, with greater recognition of environmental influences on development along with increased ability to distinguish between physical and behavioral characteristics (see Taylor et al. 2009).

Agreement with gender essentialism in children aged five to seven correlates with stronger adherence to gender expectations with respect to their personal preferences, although not with prescriptive stereotyping. In contrast, essentialist belief among parents correlates both with prescriptive stereotyping and greater gender-typed preferences among their children, but not with gender-typed preferences in the adults themselves (Meyer and Gelman 2016). Starting at age seven, if young children are taught about genetics they will apply genetic explanations to human traits, but otherwise they attribute traits to choice and environmental factors. In contrast, older children and adults tend to offer genetic explanations for at least some traits, with adults applying these explanations to a wider range of traits compared to children and using them with respect to two traits children explain only in terms of choice or environment: intelligence and athletic ability (Meyer et al. 2020). These trends suggest that genetic essentialism is learned later as children internalize the genetic-determinist beliefs circulating around them.

In-group race essentialism tends to emerge at a younger age among children from minoritized populations than the dominant group, suggesting that the experience of discrimination renders race a salient identity characteristic early in life. In an experiment involving monolingual Black and White children aged 5–6 and 9–10 years in the Chicago area, each child was shown a series of images of either White or Black children, accompanied by a voice clip in either English or French (Kinzler and Dautel 2012). For each picture, the children were shown an image of an adult that matched the pictured child's race but not language, or the reverse, and asked which of the two adults the child would grow up to resemble. White children in the 5–6 age group chose the language match more frequently than could be expected by chance, whereas White children in the 9–10 age group chose the race match. Black children in both the 5–6-year and 9–10-year age groups chose the race match, indicating earlier perception of race in Black than White children.

Children apply essentialist reasoning to their own and other groups in different ways depending on the social grid in which they grow up (Diesendruck 2021). Children from minoritized populations often express more essentialist views about their own group than the dominant group, such as five-year-old Black American and Israeli Muslim Arab children in comparison to White Americans and Israeli secular Jews, respectively. In contrast, children from dominant groups may see more uniformity and biological causation of behavior in out-groups than their in-group, as in the case of five- and eight-year-old nonimmigrant Germans and secular Israeli Jews vis-à-vis immigrant Germans and Israeli Muslim Arabs, respectively. Once children become aware of their own in-group membership, they also become aware of between-group differences in power and status and to favor the dominant group, as shown by research involving racially diverse South African four- and five-year-olds.

Although in-group empathic bias appears early in childhood and grows over time, its strength varies with the quality of relationships between groups. Positive perceptions of out-groups may attenuate in-group bias, as shown by a study comparing children and adults (Tompkins et al. 2023). While all age groups expected people to feel empathy toward their in-group in the face of both misfortune and good fortune, the degree of in-group bias depended on whether there were positive or negative relationships between groups. In addition, there was a marked age-group difference in participants' expectations of people's likely response to a misfortune experienced by an out-group. Young children (3–5 years) thought people would feel empathy toward the out-group no matter the tenor of intergroup relationships, whereas adults and older children (9–12 years) expected people to take pleasure in another group's misfortune in a context of intergroup animosity. These results suggest that positive relationships between groups can reduce in-group sentiment, and that experience in later childhood and adulthood causes the early developing empathic bias to take a bitter turn with respect to out-groups viewed negatively.

An experiment in which Jewish six-year-olds in Israeli secular schools made drawings and completed an association task illustrates how exposure to essentialist beliefs generates increased intergroup avoidance and stronger racial/ethnic essentialism (Diesendruck and Menahem 2015). The children were presented one of three stories. The Ethnic Essentialism story about the lives of an Arab and Jewish boy conveyed essentialist ideas about stable, discontinuous groups and internal homogeneity. The Ethnic Mention story included the same number of ethnically coded characters as the first story but described an unidentified boy encountering people in a park as he looked for his dog. The Animal Essentialism story replaced the boys in the first story with a giraffe and elephant. Afterward, the students drew pictures and completed a computerized task that involved assigning evaluative words to pictures of Jewish and Arab men wearing attire that marked them ethnically. Compared to the children who read the second or third story, those in the Ethnic Essentialism condition drew a picture that spaced a Jewish and Arab person farther apart and were quicker to associate pictures of Jewish men with positive evaluative words. The latter effect was especially pronounced among boys.

Starting in middle school, genetic essentialism intensifies under the influence of formal instruction as well as media and other sources of genetics/genomics information. Increased essentialist reasoning in adolescents is rooted in the same motivated cognition that fuels stereotyping and belief in genetic causes of socioeconomic inequalities in adults (see Donovan 2022; Donovan 2024b). It is likely to be fueled by the need for in-group belonging and trust, for the zeal with which young children engage in imitation of in-group models along with in-group policing of behavioral norms only grows in adolescence (Diesendruck 2021). To illustrate, the sense of belonging granted to boys in math and science courses by stereotypes about innate male ability leads to greater numbers of boys in these classrooms, whereas girls tend to predominate in courses culturally associated with femininity. Students note the gender composition of their classrooms and may then attribute the proportions to essential differences between groups (see Cheryan et al. 2025). Biology instruction on genetics and sex determination offers a ready, if mistaken, explanation, as shown in the following chapters.

By the time students enter university, they have deeply rooted essentialist preconceptions that become misconceptions if not altered by the assimilation of new knowledge (Stern and Kampourakis 2017). A study involving first-year biology majors examined the prevalence of essentialist and teleological thinking through misconception statements to which students could agree or disagree and then explain why (Stern et al. 2018).[17] The study defined essentialism the same way as this chapter, and teleology as a form of intuitive thinking that answers the question of why a trait exists with an explanation that involves a purposeful agent with an objective in mind.[18] The researchers presented students three essentialist and three teleological misconception statements, such as "Species can change during evolution, but some characteristics always remain the same" and "Animals develop protective camouflage in order to avoid their predators," respectively. Nearly all the students (97%) agreed with at least one of the six statements. Half (49%) agreed with at least three. There was no relationship between agreement with essentialist and teleological misconceptions, or the written explanations for the reasoning behind them, reflecting the contrasting forms of reasoning between the two: essentialism presupposes genetic fixity, whereas design teleology depends on mutability. The authors note that Coley and Tanner (2015) found similarly high rates of agreement with essentialist, teleological, and anthropocentric misconceptions: among biology majors, 93% agreed with at least one misconception, while among non-majors the proportion was 98%.[19]

Essentialist reasoning is less common where people interact with others in a racially diverse environment, such as school or military service.

The first of three studies by Tawa (2016) found that high school students who attend racially diverse schools are less likely to believe that race is biological, except for White students of lower socioeconomic status. The second study confirmed that agreement with race essentialism correlates with greater avoidance of people of other races. The third demonstrated that a workshop can reduce race essentialism after just one day, especially among students who express the most out-group avoidance initially. The effect remained after six weeks.

Attending university often brings students into a more diverse environment than their high school setting, resulting in a reduction in essentialist thought. Comparing data collected at the start and end of the academic year at the University of Hawai'i at Mānoa, Pauker et al. (2018) found that White first-year students from the continental US responded to increased social contact with students of other races with a decline in race essentialism. The shift was accompanied by an increase in both egalitarian views and cognitive flexibility.

Prolonged collaborative interracial interaction through military service results in reduced intergroup race avoidance and hostility, as shown by an analysis of 100–1,200 responses to each of numerous items in General Social Survey interviews from 1978 to 2021 (Green and Hyman-Metzger 2024). The study compared White men drafted for service in the Vietnam War to men who were required to sign up for the lottery but not called up for service. The second group did not show change in either racial attitudes or political orientation over the decades. The first group remained constant with respect to overall political orientation but exhibited a decline in negative opinions about Black people, avoidance of interracial contact, and agreement with discriminatory laws such as those against interracial marriage, as well as a modest increase in support for policies and programs to address racial disparities. The men who had served were more likely to have Black friends and to interact regularly with Black people in their neighborhoods, houses of worship, and work. The effect was strongest early on and declined gradually over time. Like the other studies described above, the analysis demonstrates the responsiveness of essentialist reasoning to experience and social context, despite its roots in early psychological biases.

## Social impacts of essentialism

Essentialist reasoning has such a powerful impact on perception that it can cause the brain not to recognize certain people as human. This effect has been demonstrated by functional magnetic resonance imaging that measures neural activity. Participants were shown photographs of people representing various social groups defined as warm and competent (the participant's in-group); warm but not competent; competent but not warm; or neither warm nor competent. The study looked for activity in the medial prefrontal cortex, which is involved in social cognition and activated at the sight of a human being. A control group looked at pictures of objects and verified that the sight of objects does not activate this area of the brain. In the experimental group, the brain lit up with recognition of a human being in all cases except individuals in the category of neither warm nor competent, such as homeless people and drug addicts. Analysis of the emotions evoked by pictures showed that people feel positive emotions toward the in-group and ambivalent ones regarding out-groups, especially envy toward the rich and powerful (competent but lacking in warmth) and scorn toward the elderly and ill (warm but lacking in competence). Unalloyed disgust is reserved for extreme out-groups, the sight of which correspondingly activates the insula and amygdala rather than the medial prefrontal cortex (Harris and Fiske 2006).

Based on its ubiquity as a system for categorizing social groupings, Fiske et al. (2007) argue that the warm/competent evaluation scheme could represent a universal pattern of social cognition. A study on perceptions of women provides support for the idea, for university students saw homemakers as warm but not competent, and working mothers as competent but not warm (Cuddy et al. 2004). When working women become mothers, their perceived competence declines while perceived warmth increases. In contrast, working men who become fathers lose nothing in terms of competence and gain in terms of warmth. Given the relationship between perceived competence and desirability as a worker, this conceptual scheme helps to explain discrimination against working mothers in terms of hiring, training, and job promotion. Separately, a meta-analysis of five studies indicates that undergoing pregnancy places women in a category perceived as more real, natural, and permanent than either fathers or mothers who did not give birth, such that only mothers who have undergone pregnancy are assumed to have trouble managing work and family (Park et al. 2015).

A classroom experiment devised by Jane Elliott (2018) a half-century ago unveiled the process by which essentialism leads to discrimination and prejudice. Elliott divided her Iowa third graders by eye color, offering one group praise and special privileges while denigrating and imposing limitations on the other. The first group acted upon its superiority by becoming "malicious, provocative little brown-eyed despots" and turning into much more active and effective students. The second group switched from happy, engaged learners to timid, dull, disinterested children afraid of their brown-eyed peers. The effects were easily reversed by switching the order of meanings and encouragement versus discouragement. Elliott repeated the exercise many times over several successive decades with groups ranging in age from 6 to 78 years, always with the same results.

Elliott concludes that discrimination causes prejudice – not the other way around – by way of a simple mechanism: a physical characteristic is chosen to mark certain human beings as belonging to a group; those with the characteristic are defined as inferior; expectations for that group are lowered; the group's performance falls in line with the lowered expectations; and then, their reduced performance is taken as proof of inferiority. Her experiment highlights the crucial role of the dominant group's control over power and resources, as modeled, in this case, by the teacher. It reveals the implausibility of "reverse" racism and sexism, given that disadvantaged groups lack the necessary economic and political tools (see DiAngelo 2018; West 2025).

For nondominant groups, daily life involves encounters with discrimination through all manner of experiences, such as buying a car, seeking medical care, walking around an unfamiliar neighborhood, or applying for insurance or a bank loan. At every age level, equally qualified women and ethnic minority candidates receive fewer call-backs and job offers than White men. While these disparities are resistant to diversity training, more equitable hiring emerges in response to training that emphasizes techniques for assessing work-related skills and building effective teams – a result that highlights the impact of essentialism on the way untrained managers interpret the suitability of job candidates (see Arslan et al. 2025; West 2025). Nondominant groups experience greater threats to physical autonomy in the form of house and car searches, arrests, and incarceration. They are even stopped more often for driving over the speed limit. A conservative estimate of the effect of discrimination based on Lyft data is that, at the same level of speeding, minority drivers (Black, Hispanic, Asian and Pacific Islander) are one-fourth to one-third more likely to be given a citation and are charged one-fourth to one-third higher fines than White drivers (Aggarwal et al. 2025).[20]

Not only are nondominant groups treated more harshly than Whites by law enforcement and the judicial system – with Black teens and men the most disproportionately targeted and sentenced (see Engen et al. 2002) – but criminal behavior is interpreted in essentialist terms. Court officials tend to attribute crime committed by White adolescents to external forces such as peers or a single-parent home, whereas crime by Black and Latino adolescents is chalked up to internal factors such as an undeveloped sense of remorse or an inborn propensity to delinquency (Bridges and Steen 1998).

Discrimination stemming from biological essentialism has wide-ranging negative effects on the material wellbeing, health, and physical liberty and safety of nondominant groups. Racial inequalities in housing, geographical/environmental quality, nutrition, education, physical safety, and economic wellbeing have multiple impacts on human bodies which, as seen in the following chapters, may be transmitted epigenetically from one generation to the next. Direct impacts result from unequal exposure to toxins, infectious diseases, and trauma, as well as restricted access to healthful food and the time and place for physical activity. In addition, chronic psychosocial stress contributes to racial health disparities at every level of socioeconomic status (Williams et al. 2019). Inequality and the experience of racism negatively affect physical health (low birth weight, cardiovascular disease, cancer, metabolic disorders), mental health (depression, anxiety, substance abuse), and many health indicators (heart disease risk factors, hypertension, cortisol and other markers of stress, immune and nervous system markers). Disparities pertain to all sides of illness, including onset, severity, and outcome, and to all aspects of medical care, including access, diagnosis, and treatment.

Psychosocial stress results in the same signals of accelerated aging as factors such as inactivity, smoking, and overweight/obesity, thanks to continuous activation of the hypothalamic-pituitary axis, with the release of glucocorticoid ("stress") hormones such as cortisol. The excess wear and tear manifests itself as higher allostatic load, a measure combining anthropometric indicators and indices of inflammation and cardiovascular functioning (see Murkey et al. 2022). Stress generates free radical molecules that increase oxidative stress inside cells, including those of the immune system, and shortens the telomeres at either end of the chromosomes. These clusters of short DNA repeats protect chromosomes from fusing or scrambling; they shrink with each cell replication until the cell can no longer divide and either self-destructs or enters a latent state. Psychosocial stress also suppresses the enzyme telomerase, which attaches DNA segments to telomeres. Inactivation of telomerase is linked to faster aging, separately from telomere length. Accelerated aging, in turn, correlates with higher all-cause mortality due to increased risk of infectious diseases, chronic conditions including cardiovascular disease and non-insulin-dependent diabetes, and diseases of older age such as Alzheimer's. As a result of these processes, Black Americans age faster than European Americans, similarly to mothers of chronically ill children and people who have experienced trauma or abuse.

The effects of psychosocial stress are evident from an analysis of longevity, educational attainment, and income among Black American men who "passed" as White before their death between 1975 and 2005 (Noghanibehambari and Fletcher 2025).[21] These men lived 9.4 months longer, had 2.2 more years of education, and earned 26% higher incomes than their male siblings who did not change their racial identity. Based on the increase in education, the longevity benefit should have been 1.4 months longer than observed, a deficit likely due to increased emotional stress and reduced social and family contact associated with passing.

**Essentialism and interpretations of disease**

Race essentialism not only contributes to health disparities but also distorts how they are interpreted and the way disease is treated (see Lujan and DiCarlo 2024). In the 18th and 19th centuries, myths about the greater emotionality, sensuality, and physical strength of Africans than Europeans were endorsed by mainstream scientists, who bought into an evolutionary schema that placed the former closer to nature. These ideas justified forced labor in extreme heat on swampy land infested with malaria-carrying mosquitoes, along with the use of enslaved people as experimental subjects in surgeries without anesthesia (see Williams 2009). They allowed slaveholders including US president Thomas Jefferson to reconcile their ownership and exploitation of human beings with their democratic political ideals (see DiAngelo 2018). In the 20th century, essentialist ideas about racially divergent nervous systems informed the US Public Health Service's Tuskegee study on untreated syphilis (1932–1972) and its decision to withhold penicillin from hundreds of Black men after it became widely available half-way through the project (Crenner 2012).

The alleged emotionality of people with African ancestry undergirds the notion that Black people are inherently predisposed to high blood pressure. Although the idea seems to be supported by the higher rates of hypertension among darker-skinned people in many countries, racism is a sufficient explanation. To illustrate, in Puerto Rico average blood pressure increases in step with socioeconomic level among people culturally classified as Black but decreases among those defined as White or intermediate. That is, instead of following the typical pattern of reduced stress and lower blood pressure at higher socioeconomic levels, Black people experience ever-greater psychosocial stress due to increased racism at higher levels of socioeconomic status (Gravlee et al. 2005).

Essentialist myths about African ancestry drive disparities in the way pain is managed in Black patients, who are less likely to be prescribed pain medication and are given lower doses than White patients. To assess the impact of these myths, the first of a pair of studies had 92 White US-born laypeople rate pain in a gender-matched Black and White patient across 18 scenarios (Hoffman et al. 2016). Ratings were compared to participants' level of agreement with 15 statements about alleged characteristics of Black people, such as thicker skin, slower aging, and differences in nerves and blood. Three-quarters of participants expressed agreement with at least one of the false statements (rated as possibly, probably, or definitely true). Those who expressed greater endorsement of false beliefs gave lower pain assessments for Black patients and higher pain assessments for White patients, compared to people who expressed less agreement. In the second study, 222 US-born White medical residents and students in years one to three of medical school rated their level of agreement with the 15 statements. Half of this group expressed agreement with at least one of the false beliefs. Agreement with false beliefs corresponded to lower pain ratings for mock medical cases involving Black patients and less accurate treatment recommendations.[22] Lack of agreement with the false statements correlated with higher pain ratings for Black than White patients, but no effect on treatment recommendations.

Similarly, cultural beliefs and social inequalities generate disparities in health and medical care between women and men that result in different exposures, treatments, and outcomes of the same diseases (Krieger 2024). Dissimilar treatment by health professionals that works to women's disadvantage has been documented across many conditions, including cardiovascular diseases, colorectal cancer, Parkinson's disease, and kidney disease. Dissimilar treatment unfavorable to men occurs less frequently and over a smaller range of conditions, including migraine headaches and mental health disorders such as depression (Risberg et al. 2009).

These differences recapitulate traditional cultural beliefs about the essence of femininity and masculinity that tie women's physical and mental health to their reproductive system, and that define women as more emotional than men. In the 19th and early 20th centuries, surgeons removed women's ovaries and/or uterus to treat psychological disorders. Until early this century, hormone replacement theory was routinely prescribed to treat allegedly inevitable symptoms of menopause such as depression and temperature dysregulation (see Chapter 5), and, until disproven by the Women's Health Initiative, to prevent cardiovascular and other chronic diseases and thereby justify the known increase in cancer (Rossouw et al. 2002).[23] Whether dismissed as whining, or underappreciated because women manage to endure childbirth, women's pain is widely discounted or ignored across medical conditions, from asthma to injuries (see Jackson 2021). This bias causes medical practitioners to routinely withhold procedural analgesia during IUD insertion (Tarrant et al. 2025) and to fail to refer women with chest pain to a cardiologist at the same rate as men (Clerc Liaudat et al. 2018).[24] Examples such as these illustrate how essentialist assumptions about human biology converge with the effects of discrimination to unevenly distribute health risks and outcomes across social groups.

## Essentialism and education

The essentialist idea that equates emotion with women and people of African ancestry assigns its culturally defined opposite, rationality/intelligence, to White men. This schema is evident in different groups' unequal representation in many educational and professional fields, including science, which is strongly associated with both intelligence and masculinity. As shown in Chapter 4, textbooks, teachers, and children throughout the world envision scientists predominantly as White men. Internalization of the cultural association between science and masculine brilliance begins even before elementary school, causing young girls to eschew academic activities they perceive as beyond their gender-defined ability (Wang et al. 2025). At equivalent ability levels as measured by test scores, girls underestimate their own career-related proficiency whereas boys overestimate theirs. The gap is especially wide with respect to math-heavy fields (Correll 2004).

The discrimination-induced stress response described above impacts the brain and consequently dampens performance on standardized tests and in the classroom, as Elliott found (see Pennington et al. 2016). Mechanisms include increased vigilance, which distracts the mind, obstructs thought, and increases frustration. Anxiety raises blood pressure and heart rate, with increased sweating. The brain secretes cortisol, which diverts activity from cognitive and memory centers to social and emotional areas. Negative emotions suppress the brain's pleasure centers, reducing levels of neurotransmitters such as dopamine and serotonin that stimulate cognitive functions and favor learning and performance. Experiments show that all that is needed for stereotypes to be unconsciously activated before a test is a poster on the wall or a pre-test questionnaire that asks about identity characteristics. Women perform worse and men better when triggered by questions about gender before taking a math test; Asian American perform better when asked about ethnicity but worse if asked about gender (Shih et al. 1999). As Elliott's model predicts, standardized test scores diminished by stereotype threat are taken as proof of the affected group's allegedly innate inferiority (see Steele 2011).

Comparing across countries, the degree of gender inequality and the strength of stereotypes about innate math and science ability correlate with the size of achievement gaps in these subjects at all levels from primary school to university, despite girls' advantage in reading tests and their better grades across all subjects (see Chapter 5). Stereotypes about male science ability also correlate with gender gaps in undergraduate and graduate degrees awarded. Where innatism is relatively weak, such as veterinary and biological sciences, numbers are balanced or tipped in favor of women. In contrast, men predominate in fields in which these beliefs remain stronger, such as physics, Earth sciences, engineering, and computer science. Factors that dampen women's participation in the latter fields include reduced sense of belonging due to male-oriented disciplinary cultures, lower self-efficacy, lesser early experience with the subject matter, and, in many countries, exclusion from academic and professional opportunities (Cheryan et al. 2025). These forces feed into the belief that women are less suited to science, even where they are demonstrably not. In US undergraduate courses, women outperform men in both physical and biological science courses, yet their peers continue to treat them as if they were less capable – even in biological science classrooms in which women outnumber men (Bloodhart et al. 2020).

The number of degrees awarded to non-White students likewise is out of balance with their proportion in the population overall. For instance, in the US, Black, Hispanic, and indigenous American 18–34-year-olds earn a lower proportion of STEM degrees at all levels than their 37% share of the population in their age group: 26% of undergraduate, 24% of master's, and 16% of doctoral degrees (U.S. National Science Foundation 2023). In the US, not only do women and individuals from underrepresented communities account for a significantly smaller proportion of the scientific workforce than their proportion of the population, but they are more heavily represented in lower- and middle-level positions (U.S. National Science Foundation 2023). In addition to barriers to equal participation, the generally more innovative achievements by scientists from underrepresented communities receive less recognition and result in less career advancement than they warrant (see Hofstra et al. 2020). These obstacles hinder the growth of scientific knowledge, for diversity widens the pool of perspectives and fosters innovation in all fields, including science (Kerkhoven et al. 2016).

## Histories of essentialist thought

Essentialism is nothing new, nor is its utility for justifying social inequality. In the ancient world, citizens and enslaved people, men and women, were considered fundamentally unalike, unequal, and endowed with dissimilar rights and roles. This idea fit within a system of natural classification that defined all creatures as separately constituted, as articulated by Aristotle in *Historia Animalium* (ca. 300 BCE), in the form of a *Scala Naturae* (Ladder of Nature, or Great Chain of Being). At the bottom was pure matter, without spirit: earth, water, air, and fire. Above that, worms and insects. Then shellfish, snakes, fish, birds, mammals, and humans. At the top were gods and demigods, spirit without matter. The Christian version replaced the gods with the Trinity, as in the copperplate engraving by Franciscan friar Diego de Valadés (1579). Below the Holy Spirit, the supreme White male God, and the Son of God are angels, then humans, lower creatures, plants and earth, and an underworld of spirits.

While the Christian origin story defines Woman as created from Man (Adam's rib) to serve as his companion, it does not explicitly relegate different races to an analogous subordinate position. Graves (2015) explains that in the 16th and 17th centuries, Christian writers such as Paracelsus,

Giordano Bruno, and Isaac le Peyrère nonetheless justified slavery and territorial conquest with the claim that God created different races for different parts of the world. In the US, 19th-century authors such as Josiah Priest interpreted the Bible as saying that Noah's curse on Canaan – son of his dark-skinned son Ham – gave rise to sub-Saharan African populations with character traits that destined them for slavery, whereas his light-skinned son, Japheth, was the originator of Europeans with the superior traits of masters. Graves notes that echoes of these Bible-based ideas about humankind's polygenic origins persist to the present day in Christian texts such as those published by Kingdom Identity Ministries.

Beginning in the 18th century, the Euro-American scientific and philosophical community equated humanness with the capacity for reason as exemplified by "civilized" Man. This definition cast women and non-White peoples as naturally deficient, less than human, or even nonhuman. According to Carl Linnaeus (Linné 1802), humans came in four varieties: *H. americani*, *H. europei*, *H. asiatici*, *H. afri*, plus *H. monstrosi* (with six subtypes). In addition to a particular complexion and set of physical features, each had a temperament corresponding to one of the four humors of classical medicine (yellow bile, blood, black bile, phlegm), and a system of behavioral control: choleric, stubborn Americans ruled by customs; sanguine, clever Europeans governed by laws; melancholic, proud Asians controlled by opinions; phlegmatic, unreliable Africans governed by caprice. The 18th-century German physician and naturalist, Johann Blumenbach (1795), argued that the "Caucasian" race (named for the mountain range that bisects Eurasia) was the original, finest human race, descended from Noah after his ship crashed against Mount Ararat in present-day Turkey. The Caucasian race sat alone at the top of the Great Chain of Being. All others were a "degeneration": Mongolians, Malayans (Oceania and Southeast Asian islands), indigenous Americans, and, most degraded in Blumenbach's view, Ethiopians (all of Africa).

During the first half of the 19th century, the American physician and naturalist, Samuel George Morton, sought to prove that Europeans had larger brains than other "races" by measuring the volume of ancient and modern skulls from Blumenbach's same groups, which Morton called White Races or Europeans, Malays, Mongolians, Negroes, and Americans. Morton collected skulls and filled them with seeds and then bullets for greater precision. He arranged 15 samples in descending order from the "Teutonic Group" of the "Modern White Races" at 92 cubic inches to Hottentots and Australians (tied at 75) (Nott and Gliddon 1854). Sample sizes were highly unbalanced and drawn from dissimilar sources, with overrepresentation of indigenous American skulls due most likely to the inability of these communities to stop him from taking them (Graves 2015). A re-analysis of the data grouped into the five categories plus ancient Europeans by Stephen Jay Gould (1981) moved the latter to second-to-bottom and Mongolians to top position shared with modern Europeans. Gould's analysis has been criticized, but, as he thoroughly documents, no analysis of cranial volume can shed light on population differences in cognitive ability, given that, within the normal range, intelligence is unrelated to brain size.

In the mid to late 19th century, some scientists joined religious scholars in backing the polygenic origin story. The 19th-century Swiss-American geologist-biologist, Louis Agassiz, asserted that the world's creatures, including distinct varieties of humans, were divided into eight separate geographical "realms" (Nott and Gliddon 1854). Agassiz and Morton's student, Josiah Nott, gave public and scientific presentations on the subject, including a lecture by Agassiz to the American Association for the Advancement of Science in 1850 and numerous speeches by Nott in defense of slavery (Graves 2015).

The US politician/attorney/anthropologist, Lewis Henry Morgan (1877), placed human societies along a progression of stages from savagery to barbarism and finally civilization. The science of the day considered these stages to be legible not only in brain size but also numerous physical characteristics such as facial features, hair, toes, cranial protrusions, and Paul Broca's ratio of radius-to-humerus length (see Horn 2015; Whitaker 2000). Drawings ranked individual exemplars from nonhuman primates to Brown and Black men, and finally White men. White women's brain size and complexity, and presumed moral and intellectual quality, were considered intermediate between White men's and those of apes and male primitives. Scientists seemed not to notice that brains varied among White men, including eminent ones with small heads. Gould (1981:90) juxtaposes a 1903 schematic by the American anatomist E.A. Spitska of brains along a scale from gorilla to "Bushwoman" to mathematician K.F. Gauss, against a drawing in the same publication that compares the brains of three eminent men, one of which closely resembles the "Bushwoman's" in size and shape.

The 19th-century science of craniometry focused on the roundness of the skull as the indicator of evolutionary grade, with longer skulls housing a more prominent cerebral cortex that placed them in a higher position.[25] The cephalic (or cranial) index attributed to the Swedish anatomist, Anders Retzius, divides breadth by length, such that smaller values indicate long-headedness (dolichocephaly) while larger ones signify round-headedness (brachycephaly).[26] Dolichocephalic skulls were common among Indo-Europeans; brachycephalic ones among Paleolithic samples. Among Indo-European populations, those of northern European countries such as Sweden, England, and Germany had higher proportions of dolichocephalic skulls, whereas southern European countries had higher proportions of brachycephalic ones (Gould 1981).[27]

While some comparisons fit the evolutionary progression, others did not, including subpopulations within European countries and long-headed Australians and Africans, as shown by the American sociologist William Ripley (1899). Ripley questioned the connection between head shape and cultural backwardness, drawing upon craniometric data from scientists including the German-American physicist-turned-anthropologist Franz Boas. Boas was interested in the roles of heredity and environment in head-shape variation, leading to meticulous analyses of subpopulations in Europe and North America (see Boas and Boas 1913). His measurements on the skulls of thousands of immigrants to the US revealed that a population's average skull shape could vary considerably over time (Boas 1912). The immigrants included Bohemians and Sicilians, Slovaks and Hungarians – groups then considered unique races with distinctive mental and behavioral traits. Boas's comparisons of these measurements to those of US-born children of immigrants showed that a population's average head shape could sway from brachycephaly to dolichocephaly, or vice versa, in a single generation. The longer the mother had lived in the US, the greater the change in the next generation. The study cast doubt on the usefulness of head shape as an indicator of racial/evolutionary status and set Boas on a path of arguing against biological, geographical, and every other form of determinism (see Boas 1932).[28] He was joined by the British-American anthropologist Ashley Montagu (1942), who showed that modern evolutionary theory and genetics knowledge lent no scientific support to the biological race concept.

As many mainstream scholars in anthropology and biology turned away from biological essentialism, the race concept gained ground in US law and society (see DiAngelo 2018; Jackson and Depew 2017). Until mid-century, the US Eugenics Record Office at Cold Spring Harbor maintained a register of traits, including intelligence, feeblemindedness, and chess ability, under the presumption that every behavioral and cognitive feature had a single-gene origin (Graves 2015). Not just dictatorships but also democracies implemented eugenic policies aimed at improving the biological quality of their populations through birth incentives for people with valued traits and forced sterilization and other measures to curtail reproduction among those with disvalued characteristics (see Teicher 2020; Whitaker 2000). Although eugenics is no longer a mainstream science, these ideas and some of the same initiatives have lately resurfaced in Europe and the US in response to immigration from countries with higher birth rates perceived now, as in the past, as biologically threatening to national populations.

The postwar retreat of scientific racism was short-lived. In a treatise that continues to reverberate through hereditarian writings about the uselessness or even counter-productiveness of efforts to address racial disparities, educational psychologist Arthur Jensen (1969) argued that intelligence is unaffected by postnatal influences, aside from extreme isolation. Jensen mischaracterized heritability statistics (see Chapter 2) as expressing the portion of parent-offspring correlation in intelligence, educational attainment, occupational status, and income that is genetically determined, to assert that racial achievement gaps in education were brittle rather than dependent on social factors.

Scientists immediately contested Jensen's argument (see Graves 2015; Lala and Feldman 2024; Sussman 2016). In the early 1970s, Luca Cavalli-Sforza, Marcus Feldman, and Richard Lewontin showed that heritability scores for intelligence and academic and economic outcomes can be accounted for by non-genetic transmission from parents to children (Cavalli-Sforza and Feldman 1973; Feldman and Lewontin 1975; Lewontin 1970). As a result, high heritability does not mean these traits are resistant to change through policies and programs. On the contrary, scores on intelligence tests vary widely across time scales impossible for genetic change (see Chapter 2), indicating the importance of environmental, sociocultural, and family variables.

Despite continued opposition from the scientific community, hereditarians have persisted in manipulating biological principles and using questionable data and statistical methods to assert that racial disparities are innate and immutable. Based on an inappropriate application of r-K selection to groups within the same species as opposed to different species with vastly dissimilar reproductive rates, parental investment, and life spans, psychologist J. Philippe Rushton (1995) argued that the more challenging climatic conditions faced by populations outside Africa acted as a selective force that left the descendants of Europeans ("Caucasoids") and East Asians ("Mongoloids" or "Orientals") intellectually, morally, and behaviorally superior to people of African origin ("Negroids"). Gene-culture coevolution, which concerns the spread of physical traits in particular populations in relation to factors such as diet or exposure to infectious diseases (see Chapter 3), has been used by authors such as journalist Nicholas Wade (2015) to argue that culture acts as a selective force on cognitive traits as well. To illustrate, anthropologists Gregory Cochrane and Henry Harpending (2009) attribute high intelligence in Ashkenazi Jews to selective pressure exerted by European societies that placed Jews in role of financiers. Political scientist Charles Murray (2020) draws on genomics to support the notions presented with Richard Herrnstein decades ago (1994) that intelligence is impervious to training and that achievement disparities derive from evolved differences between races. Murray argues that races are real and differ biologically and, in time, will be shown to differ in cognitively as well; that gender is the same as sex; and that social class has nothing to do with inherited privilege.

Economist Gregory Clark (2023) advances Jensen's argument through the claim that social mobility has remained unchanged in England for the last four centuries despite the expansion of public education, due to "genetic nurture" whereby the genes of elite parents (joined by assortative mating) result in a favorable social environment that enhances child outcomes. Clark's study generated a constant rate of intergenerational difference in correlated social outcomes among relatives up to the fourth degree in a sample of approximately 420,000 people. However, the statistical analysis did not include wealth as a variable and has since been refuted. As Lala and Feldman (2024) point out, even if it were correct, the finding would be more plausibly explained by the tendency of elite families to advance their children and preserve inherited wealth. In fact, survey data on education and occupation of respondents and their parents, provided by 176,000 people born between 1920 and 1979 in 40 countries, indicate that education does correlate with increased intergenerational mobility, notwithstanding the resilience of this protective tendency in the face of expanding access to educational opportunities (Van de Werfhorst 2024).

Direct genetic explanations for economic differences between populations, and especially the predominance of Eurasian societies, include the argument by evolutionary geneticists Evans et al. (2006) that a single gene, *MCPH1* (microcephalin), is the reason agriculture and cities emerged in the Middle East and Europe instead of Africa. The gene plays a role in regulating brain development and size, and which spread rapidly under positive selection after introduction into *H. sapiens* from an archaic *Homo* lineage 37,000 years ago in Eurasia. According to economists Quamrul Ashraf and Oded Galor (2013), too much and too little genetic diversity have been "detrimental" to the economic development of Africa and indigenous America, respectively, whereas middle values have been "conducive" to development in Eurasia – measured as per-capita income in the industrial era and population density beforehand. These ideas not only discount the political, economic, and technological achievements of African and indigenous American societies. They contradict biological principles, for there is no theory or evidence to suggest that specific genes or the amount of genetic diversity has anything to do with economic development (Lala and Feldman 2024).

Just as scientific racism has served as the handmaiden of territorial expansion, slavery, and colonization, scientific sexism has facilitated the legal and social subordination of women. Scientific studies supporting restricted rights and roles for women have tended to proliferate during times of social unrest or labor competition. For instance, 19th-century studies of brain size justified women's exclusion from political participation and their institutionalized economic dependence on men. In the 20th century, scientific studies that supported restrictions on women waxed and waned in line with the degree of labor competition. During periods in which women's paid employment was threatening to men's, studies expressed the essentialist association of women with childrearing and claimed that women's work outside the home, as well as any form of "cerebralism" including reading novels, damaged their reproductive capacities and competence as wives and mothers, with negative impacts on children's wellbeing. Opposite results were reported during the two world wars, when women's labor was needed (Martin 2001; Whitaker 2000).[29] That men's work outside the home is not similarly framed in relation to reproduction and parenting reflects the continued force of essentialist ideas about natural gender roles established in the deep past, discussed in Chapter 6.

The resurgence of scientific racism and sexism over the past few decades corresponds with a rise in socioeconomic inequality within and between countries, which has brought an unprecedented accumulation of wealth by individuals and corporations. All 15 of the top billionaires in the Bloomberg Billionaire Index are White; 14 are men (the one woman is a member of the Walmart family); and all are American or European (13 US; 1 France; 1 Spain) (9 April 2025). Any one of these people's fortunes is greater than the entire gross domestic product of dozens of countries, including Hungary, Kenya, Slovakia, and Ukraine. Data from the CIA World Factbook, Fortune Global 500, and the OECD consistently show that large corporations, whose leadership is predominantly male and White, bring in more revenue than many national governments, including those of wealthy countries such as Sweden and Belgium.

The US stands out as especially affluent and unequal (and unhappy and unhealthy; see Mishra and Dash 2023) relative to other industrialized nations, with a gap between the highest- and lowest-paid private-sector workers far wider than in all other wealthy countries. Inequality between nations and US states correlates with worse health and higher mortality; lower social integration, social capital, mutual trust, and trust in institutions; lower educational performance; and higher rates of crime, fear of crime, and many other social ills (Wilkinson and Pickett 2009). This context, together with the perceived economic and social losses of White men relative to other groups, has fueled the rise in racist and sexist rhetoric discussed above.

Biological essentialism holds that it is not structures of inequality, but innate intellectual, physical, and behavioral inferiority that explain why some groups have less wealth, worse health, and lower representation in academia, business leadership, and politics and government. Essentialism renders social change pointless and unnatural and lends support to actions that buttress or augment the power and privilege of dominant groups. As seen in Chapter 7, inequality and supporting ideologies go hand in hand. However, the intertwined pair is neither permanent nor inevitable, which suggests that refuting genetic essentialism can weaken the foundations on which inequality is built. The following chapters explore the key role science education can play in advancing an anti-essentialist understanding of human variation that is ethical, accurate, and congruent with human biology.

## Notes

1 For instance, an analysis of mortality among 73,838 individuals aged 50–85 years over the decade beginning in 2010 in the US and 16 European countries (with 13,802 deaths) found a similar survival rate for the wealthiest one-fourth of Americans and the poorest one-fourth in Western and Northern Europe. In addition, while the top three quartiles of wealth in all countries correlated with better survival than the lowest quartile, in the US the distance between the top and bottom quartiles was greater than in Europe (Machado et al. 2025).
2 In the US, "blind" auditions became increasingly popular beginning in the 1970s and 1980s. By the mid-1990s, the proportion of women new hires had increased from less than 10% to 35% in the Boston and Chicago Symphony Orchestras, and to 50% in the New York Philharmonic. Among a sample of eight orchestras, the proportion of women musicians had risen from 0–12% to 20–35%. These increases occurred despite a reduction in turnover over the same period (Goldin and Rouse 2000).
3 The impact of these formats is considerable, given that people obtain most of their scientific knowledge from informal sources such as print and visual media, and park and museum exhibits (see Falk and Dierking 2010).
4 For US statistics, see the FBI, DOJ, and Anti-Defamation League websites. In 2023, three-quarters of single-bias hate crimes (11,447 of a total of 13,829) were motivated by race/ethnicity/ancestry (52.5%) or religion (22.5%). Nearly a fifth were motivated by sexual orientation (18.4%), with the rest divided among gender identity, disability, and gender in descending order (https://www.justice.gov/hatecrimes/hate-crime-statistics).
5 Donovan et al. (2024a) report that 40–70% of US adults believe that genes determine cognitive/behavioral differences between genders.
6 Unless otherwise stated, all studies discussed in this book were based in the US.
7 While recognition of genetic factors in complex diseases such as cancer and heart disease may be expected to reduce belief in the role of lifestyle factors, a survey of 1,747 British adults found that the one-third of people who recognize the first are also more likely to believe in the second. Groups more likely to recognize genetic factors in these diseases included women, older people, more educated people, and, for heart disease, those with a family history (Sanderson et al. 2011).
8 Canada.
9 A control group received a reading on an unrelated topic.
10 There was no difference in effects on essentialism between the type of ancestries identified by the tests.
11 Overall, genetics knowledge was similar between women and men, and Black and White respondents.
12 For the contrary view that people tend not to conflate bio-somatic with bio-behavioral essentialism, see Condit (2019), who argues that racism is anterior to explanations for differences between groups. However, the argument is based on focus groups and surveys, which are relatively ineffective at probing unconscious biases.
13 These associations are based on a meta-analysis of five studies.
14 This is based on a meta-analysis of eight studies.
15 Study 1, Australia; Study 2, UK.
16 The acronym stands for Biology, Health, and Environmental Education for Better Citizenship.

17 France.
18 As the authors explain, if purposeful design were correct, then both sharks and whales would have not only a hydrodynamic shape but also the gills that allow only sharks to remain underwater.
19 The French study did not include anthropocentric misconceptions, given that anthropomorphism is a form of teleological thinking.
20 The data concern over 220,000 drivers in Florida. A first analysis yielded a 33% greater likelihood of a speeding citation and a 34% higher fine; a second, 24% and 23%, respectively. The proportions are likely to understate the effects of discrimination, given that incentives for good driving on the part of Lyft drivers result in relatively few stops for speeding and relatively low speeds over the limit.
21 $n$ = 4,674.
22 The authors point out that part of the disparity in pain assessment could be due to White in-group bias or beliefs about White frailty.
23 While the 20-year Women's Health Initiative determined that hormone replacement therapy increases the risk of both cardiovascular disease and cancer, recent studies indicate that it is safe and effective for treating vasomotor symptoms in women under 60 without risk factors (see Manson et al. 2024).
24 In a study involving 672 outpatient-clinic patients in Switzerland, Clerc Liaudat et al. (2018) found a similar prevalence of chest pain between men and women but 2.5 times more referrals to a cardiologist for men than women.
25 Using an apparatus designed for the purpose, breadth was measured between two points behind the cheekbones, above the supramastoid crest of the temporal bones, and length from the space between the eyebrows above the nose (glabella) and the center of the occipital bone (occiput).
26 A cranial index over 0.8 denotes round-headedness or brachycephaly; 0.75–0.80 mesocephaly; and, below 0.75 long-headedness or dolichocephaly.
27 Broca, whose skull was brachycephalic as was typical in France, made sure to find Paleolithic dolichocephalic skulls, as well as brachycephalic heads among subgroups of Nordic populations (Gould 1981).
28 For a re-analysis of Boas's data that supports his conclusions, along with a refutation of a re-analysis that suggests head shape was not significantly affected by the US environment, see Gravlee et al. (2003).
29 For the current scientific consensus that women's work outside the home favors their and their children's health and wellbeing, at least after the child's first year of life (before which results are mixed), see Frech and Damaske (2012); Lucas-Thompson et al. (2010).

## References

Aggarwal P, Brandon A, Goldszmidt A, Holz J, List JA, et al. 2025. High-frequency location data show that race affects citations and fines for speeding. *Science* 387(6741):1397–1401.

Allen CG, Andersen B, Khoury MJ, Roberts MC. 2019. Current social media conversations about genetics and genomics in health. *Public Health Genomics* 21(1–2):93–99.

Arslan C, Change EH, Chilazi S, Bohnet I, Hauer OP. 2025. Behaviorally designed training leads to more diverse hiring. *Science* 387(6732):364–366.

Ashraf Q, Galor O. 2013. The "Out of Africa" hypothesis, human genetic diversity, and comparative economic development. *American Economic Review* 103(1):1–46.

Bailey AH, Knobe J, Newman GE. 2021. Value-based essentialism. *Journal of Experimental Psychology: General* 150(10):1994–2014.
Bloodhart B, Balgopal MM, Casper AM, Sample McMeeking LB, Fischer EV. 2020. Outperforming yet undervalued. *PLoS One* 15(6):e0234685.
Blumenbach JF, Banks J. 1795. *De generis humani varietate nativa*. Vandenhoek et Ruprecht.
Boas F. 1912. Changes in the bodily form of descendants of immigrants. *American Anthropologist* 14(3):530–562.
Boas F. 1932. The aims of anthropological research. *Science* 76(1983):605–613.
Boas F, Boas HM. 1913. The head-forms of the Italians as influenced by heredity and environment. *American Anthropologist* 15(2):163–188.
Brescoll VL, Uhlmann EL, Newman GE. 2013. The effects of system-justifying motives on endorsement of essentialist explanations for gender differences. *Journal of Personality and Social Psychology* 105(6):891–908.
Bridges GS, Steen S. 1998. Racial disparities in official assessments of juvenile offenders. *American Sociological Review* 63(4):554–570.
Castéra J, Clément P. 2014. Teachers' conceptions about the genetic determinism of human behaviour. *Science & Education* 23:417–443.
Cavalli-Sforza LL, Feldman MW. 1973. Cultural versus biological inheritance. *American Journal of Human Genetics* 25(6):618–637.
Cheryan S, Lombard EJ, Hailu F, Pham LN, Weltzien K. 2025. Global patterns of gender disparities in STEM and explanations for their persistence. *Nature Reviews Psychology* 4(1):6–19.
Christensen KD, Jayaratne TE, Roberts JS, Kardia SL, Petty EM. 2010. Understandings of basic genetics in the United States. *Public Health Genomics* 13(7–8):467–476.
Clark G. 2023. The inheritance of social status. *Proceedings of the National Academy of Sciences* 120(27):e2300926120.
Clerc Liaudat C, Vaucher P, De Francesco T, Jaunin-Stalder N, Herzig L, et al. 2018. Sex/gender bias in the management of chest pain in ambulatory care. *Women's Health*. https://doi.org/10.1177/1745506518805641
Cochrane G, Harpending H. 2009. *The 10,000–year explosion*. Basic Books.
Coley JD, Tanner K. 2015. Relations between intuitive biological thinking and biological misconceptions in biology majors and nonmajors. *CBE—Life Sciences Education* 14(1):ar8.
Condit CM. 2019. Laypeople are strategic essentialists, not genetic essentialists. *Hastings Center Report* 49:S27–S37.
Condit CM, Parrott RL, Bates BR, Bevan J, Achter PJ. 2004. Exploration of the impact of messages about genes and race on lay attitudes. *Clinical Genetics* 66(5):402–408.
Correll SJ. 2004. Constraints into preferences. *American Sociological Review* 69(1):93–113.
Crenner C. 2012. The Tuskegee Syphilis Study and the scientific concept of racial nervous resistance. *Journal of the History of Medicine and Allied Sciences* 67(2):244–280.
Cuddy AJ, Fiske ST, Glick P. 2004. When professionals become mothers, warmth doesn't cut the ice. *Journal of Social Issues* 60(4):701–718.
Dar-Nimrod I, Heine SJ. 2011. Genetic essentialism. *Psychological Bulletin* 137(5):800–818.
Dasgupta K, Iturriaga N, Panofsky A. 2024. Beyond biological essentialism. *Social Problems* 71(4):1200–1215.
DiAngelo R 2018. *White fragility*. Beacon Press.

Diesendruck G. 2021. A motivational perspective on the development of social essentialism. *Current Directions in Psychological Science* 30(1):76–81.
Diesendruck G, Menahem R. 2015. Essentialism promotes children's inter-ethnic bias. *Frontiers in Psychology* 6:1180.
Donovan BM. 2017. Learned inequality. *Journal of Research in Science Teaching* 54(3):379–411.
Donovan BM. 2022. Ending genetic essentialism through genetics education. *Human Genetics and Genomics Advances* 3:1–13.
Donovan BM, Semmens R, Keck P, Brimhall E, Busch KC, et al. 2019. Toward a more humane genetics education. *Science Education* 103(3):529–560.
Donovan BM, Syed A, Arnold SH, Lee D, Weindling M, et al. 2024a. Sex and gender essentialism in textbooks. *Science* 383(6685):822–825.
Donovan BM, Weindling M, Amemiya J, Salazar B, Lee D, et al. 2024b. Humane genomics education can reduce racism. *Science* 383(6685):818–822.
Donovan BM, Weindling M, Salazar B, Duncan A, Stuhlsatz M, et al. 2021. Genomics literacy matters. *Journal of Research in Science Teaching* 58(4):520–550.
Elliott J. 2018. It's all about ignorance. *The Cambridge Handbook of the Psychology of Prejudice* 11:253–267.
Engen RL, Steen S, Bridges GS. 2002. Racial disparities in the punishment of youth. *Social Problems* 49(2):194–220.
Evans PD, Mekel-Bobrov N, Vallender EJ, Hudson RR, Lahn BT. 2006. Evidence that the adaptive allele of the brain size gene microcephalin introgressed into Homo sapiens from an archaic Homo lineage. *Proceedings of the National Academy of Sciences* 103(48):18178–18183.
Falk JH, Dierking LD. 2010. School is not where most Americans learn most of their science. *American Scientist* 98(6):486.
Feldman MW, Lewontin RC. 1975. The heritability hang-up. *Science* 190(4220):1163–1168.
Fiske ST, Cuddy AJ, Glick P. 2007. Universal dimensions of social cognition. *Trends in Cognitive Sciences* 11(2):77–83.
Frech A, Damaske S. 2012. The relationships between mothers' work pathways and physical and mental health. *Journal of Health and Social Behavior* 53(4):396–412.
Gallup.com. 2024. Majority still credits God for humankind, but not creationism. https://news.gallup.com/poll/647594/majority-credits-god-humankind-not-creationism.aspx
Gericke N, Carver R, Castéra J, Evangelista NA, Marre CC, et al. 2017. Exploring relationships among belief in genetic determinism, genetics knowledge, and social factors. *Science & Education* 26:1223–1259.
Goldin C, Rouse C. 2000. Orchestrating impartiality. *American Economic Review* 90(4):715–741.
Gould SJ. 1981. *The mismeasure of man*. WW Norton & Company.
Graves Jr JL. 2015. Great is their sin. *The Annals of the American Academy of Political and Social Science* 661(1):24–50.
Gravlee CC, Bernard HR, Leonard WR. 2003. Boas's changes in bodily form. *American Anthropologist* 105(2):326–332.
Gravlee CG, Dressler WW, Bernard HR. 2005. Skin color, social classification, and blood pressure in southeastern Puerto Rico. *American Journal of Public Health* 95(12):2191–2197.
Green DP, Hyman-Metzger O. 2024. The Vietnam draft lottery and whites' racial attitudes. *American Political Science Review* 119(4):2011–2018.

Haran J, O'Riordan K. 2018. Public knowledge-making and the media. *European Journal of Cultural Studies* 21(6):687–706.
Harris LT, Fiske ST. 2006. Dehumanizing the lowest of the low. *Psychological Science* 17(10):847–853.
Heine SJ, Cheung BY, Schmalor A. 2019. Making sense of genetics. *Hastings Center Report* 49:S19–26.
Herrnstein R, Murray C. 1994. *The bell curve*. Free Press.
Hoffman KM, Trawalter S, Axt JR, Oliver MN. 2016. Racial bias in pain assessment and treatment recommendations, and false beliefs about biological differences between blacks and whites. *Proceedings of the National Academy of Sciences* 113(16):4296–4301.
Hofstra B, Kulkarni VV, Munoz-Najar Galvez S, He B, Jurafsky D, et al. 2020. The diversity–innovation paradox in science. *Proceedings of the National Academy of Sciences* 117(17):9284–9291.
Horn D. 2015. *The criminal body*. Routledge.
Jackson G. 2021. *Pain and prejudice*. Greystone Books Ltd.
Jackson J, Depew D. 2017. *Darwinism, democracy, and race*. Routledge.
Jensen AR. 1969. How much can we boost IQ and scholastic achievement? *Harvard Educational Review* 39:1–123.
Kampourakis K, Fux M. 2025. Ancestry inferences from DNA testing results. *History and Philosophy of the Life Sciences* 47(2):25.
Keller J. 2005. In genes we trust. *Journal of Personality and Social Psychology* 88(4):686–702.
Kerkhoven AH, Russo P, Land-Zandstra AM, Saxena A, Rodenburg FJ. 2016. Gender stereotypes in science education resources. *PLoS One* 16(11):e0165037.
King TL, Scovelle AJ, Meehl A, Milner AJ, Priest N. 2021. Gender stereotypes and biases in early childhood. *Australasian Journal of Early Childhood* 46(2):112–225.
Kinzler KD, Dautel JB. 2012. Children's essentialist reasoning about language and race. *Developmental Science*. 15(1):131–138.
Krieger N. 2024. Advancing gender transformative intersectional science for health justice. *Social Science & Medicine* 351:116151.
Lala KN, Feldman MW. 2024. Genes, culture, and scientific racism. *Proceedings of the National Academy of Sciences* 121(48):e2322874121.
Lawless B, Chen YW. 2018. An eye is an eye makes the world colorblind. In Atay A, Ashlock MZ, editors. *Millennial Culture and Communication Pedagogies*. Lexington Books. pp. 93–106.
Lee KY, Reis HT, Rogge RD. 2020. Seeing the world in pink and blue. *Sex Roles* 83(11):685–705.
Lewontin RC. 1970. Race and intelligence. *Bulletin of the Atomic Scientists* 26(3):2–8.
Linné C. 1802. *System of nature*. Turton W, translator. Lackington, Allen, and Co. p. 9.
Little ID, Koehly LM, Gunter C. 2022. Understanding changes in genetic literacy over time and in genetic research participants. *The American Journal of Human Genetics* 109(12):2141–2151.
Lo RF, Sasaki JY. 2024. Lay misperceptions of culture as "biological" and suggestions for reducing them. *Perspectives on Psychological Science* 19(1):295–311.
Lucas-Thompson RG, Goldberg WA, Prause J. 2010. Maternal work early in the lives of children and its distal associations with achievement and behavior problems. *Psychological Bulletin* 136(6):915–942.

Lujan HL, DiCarlo SE. 2024. Misunderstanding of race as biology has deep negative biological and social consequences. *Experimental Physiology* 109(8):1240–1243.
Lynch J, Bevan J, Achter P, Harris T, Condit CM. 2008. A preliminary study of how multiple exposures to messages about genetics impact on lay attitudes towards racial and genetic discrimination. *New Genetics and Society* 27(1):43–56.
Machado S, Kyriopoulos I, Orav EJ, Papanicolas I. 2025. Association between wealth and mortality in the United States and Europe. *New England Journal of Medicine* 392(13):1310–1319.
Manson JE, Crandall CJ, Rossouw JE, Chlebowski RT, Anderson GL, et al. 2024. The Women's Health Initiative randomized trials and clinical practice. *JAMA* 331(20):1748–1760.
Martin E. 2001. *The woman in the body*. Beacon Press.
McCartney J. 2023. Exposing the hazards of teaching 19th century genetic science. *Journal of Research in Science Teaching*. https://doi.org/10.1002/tea.21895
McKown C, Weinstein RS. 2003. The development and consequences of stereotype consciousness in middle childhood. *Child Development* 74(2):498–515.
Meyer M, Gelman SA. 2016. Gender essentialism in children and parents. *Sex Roles* 75:409–421.
Meyer M, Roberts SO, Jayaratne TE, Gelman SA. 2020. Children's beliefs about causes of human characteristics. *Journal of Experimental Psychology: General* 149(10):1935–1949.
Mills Shaw KR, Van Horne K, Zhang H, Boughman J. 2008. Essay contest reveals misconceptions of high school students in genetics content. *Genetics* 178(3): 1157–1168.
Mishra PP, Dash S. 2023. World happiness report. In *The Palgrave encyclopedia of global security Studies*. Springer International Publishing. pp. 1595–1599.
Montagu MFA. 1942. *Man's most dangerous myth*. Columbia University Press.
Moore DS, Kampourakis K, Gericke N. 2025. The problems with genetic essentialism, determinism, and reductionism. *Human Development* 69(3):147–157.
Morgan LH. 1877. *Ancient society*. Henry Holt & Co.
Morin-Chassé A. 2014. Public (mis) understanding of news about behavioral genetics research. *BioScience* 64(12):1170–1177.
Morin-Chassé A. 2020. Behavioral genetics, population genetics, and genetic essentialism. *Science & Education* 29(6):1595–1619.
Morton TA, Hornsey MJ, Postmes T. 2009a. Shifting ground. *British Journal of Social Psychology* 48(1):35–59.
Morton TA, Postmes T, Haslam SA, Hornsey MJ. 2009b. Theorizing gender in the face of social change. *Journal of Personality and Social Psychology* 96(3):653–664.
Murkey JA, Watkins BX, Vieira D, Boden-Albala B. 2022. Disparities in allostatic load, telomere length and chronic stress burden among African American adults. *Psychoneuroendocrinology* 140:105730.
Murray C. 2020. *The biology of gender, race, and class*. Twelve.
Nam HH, Sawyer K. 2024. Scientific supremacy. *Politics and the Life Sciences* 43(1):99–131.
Nelkin D, Lindee 2010. *The DNA mystique*. University of Michigan Press.
Noghanibehambari H, Fletcher J. 2025. Passing as White. *National Bureau of Economic Research Working Paper 33394*. http://www.nber.org/papers/w33394.
Norton HL, Quillen EE, Bigham AW, Pearson LN, Dunsworth H. 2019. Human races are not like dog breeds. *Evolution: Education and Outreach* 12(17):1–20.

Nott JC, Gliddon GR. 1854. *Types of mankind. Illustrated by selections from the inedited papers of Samuel George Morton and by additional contributions from L Agassiz, W Usher, and HS Patterson*. Lippincott, Grambo & Company.
Outram S, Graves JL, Powell J, Wolpert C, Haynie KL, et al. 2018. Genes, race, and causation. *Race and Social Problems* 10:79–90.
Park B, Banchefsky S, Reynolds EB. 2015. Psychological essentialism, gender, and parenthood. *Journal of Personality and Social Psychology* 109(6):949–967.
Parrott R, Smith RA. 2014. Defining genes using "blueprint" versus "instruction" metaphors. *Health Communication* 29(2):137–146.
Pauker K, Carpinella C, Meyers C, Young DM, Sanchez DT. 2018. The role of diversity exposure in Whites' reduction in race essentialism over time. *Social Psychological and Personality Science* 9(8):944–952.
Pearce A, Mitchell LA, Best S, Young MA, Terrill B. 2024. Publics' knowledge of, attitude to and motivation towards health-related genomics. *European Journal of Human Genetics* 32(7):747–758.
Pennington CR, Heim D, Levy AR, Larkin DT. 2016. Twenty years of stereotype threat research. *PLoS One* 11(1):e0146487.
Phelan JC, Link BG, Feldman NM. 2013. The genomic revolution and beliefs about essential racial differences. *American Sociological Review* 78(2):167–191.
Phelan JC, Link BG, Zelner S, Yang LH. 2014. Direct-to-consumer racial admixture tests and beliefs about essential racial differences. *Social Psychology Quarterly* 77(3):296–318.
Pobiner B. 2016. Accepting, understanding, teaching, and learning (human) evolution. *American Journal of Physical Anthropology* 159(S61):232–274.
Rhodes M, Mandalaywala TM. 2017. The development and developmental consequences of social essentialism. *Wiley Interdisciplinary Reviews: Cognitive Science* 8(4):e1437.
Ripley WZ. 1899. *The races of Europe*. D. Appleton.
Risberg G, Johansson EE, Hamberg K. 2009. A theoretical model for analysing gender bias in medicine. *International Journal for Equity in Health* 8:1–8.
Roberts SO, Ho AK, Rhodes M, Gelman SA. 2017. Making boundaries great again. *Personality and Social Psychology Bulletin* 43(12):1643–1658.
Rossouw JE, Anderson GL, Prentice RL, LaCroix AZ, Kooperberg C, et al. 2002. Risks and benefits of estrogen plus progestin in healthy postmenopausal women. *JAMA* 288(3):321–333.
Roth WD, Yaylacı Ş, Jaffe K, Richardson L. 2020. Do genetic ancestry tests increase racial essentialism? *PLoS One* 15(1):e0227399.
Rushton JP. 1995. *Race, evolution, and behavior*. Transaction Publishers.
Ryazanov AA, Christenfeld NJ. 2018. The strategic value of essentialism. *Social and Personality Psychology Compass* 12(1):e12370.
Sanderson SC, Waller J, Humphries SE, Wardle J. 2011. Public awareness of genetic influence on chronic disease risk. *Public Health Genomics* 14(4–5):290–297.
Sawyer K, Nam HH. 2024. The genetic essentialism of the alt-right. *Research & Politics* 11(2):20531680241252465.
Schneider SP, Smith KB, Hibbing JR. 2018. Genetic attributions. *The Journal of Politics* 80(3):1023–1027.
Shih M, Pittinsky TL, Ambady N. 1999. Stereotype susceptibility. *Psychological Science* 10(1):80–83.
Shostak S, Freese J, Link BG, Phelan JC. 2009. The politics of the gene. *Social Psychology Quarterly* 72(1):77–93.
Steele C. 2011. *Whistling Vivaldi*. WW Norton & Company.

Stern F, Kampourakis K. 2017. Teaching for genetics literacy in the post-genomic era. *Studies in Science Education* 53(2):193–225.

Stern F, Kampourakis K, Huneault C, Silveira P, Müller A. 2018. Undergraduate biology students' teleological and essentialist misconceptions. *Education Sciences* 8(135). https://doi.org/10.3390/educsci8030135

Stern F, Kampourakis K, Müller A. 2023. "Genes for a role," "genes as essences." *Journal of Research in Science Teaching* 60(2):237–267.

Sussman RW 2016. *The myth of race*. Harvard University Press.

Tarrant L, Grills R, McLeod K. 2025. Gender bias in pain management. *Australian and New Zealand Journal of Obstetrics and Gynaecology*. https://doi.org/10.1111/ajo.70010

Tawa J. 2016. Belief in race as biological. *Race and Social Problems* 8:244–255.

Taylor MG, Rhodes M, Gelman SA. 2009. Boys will be boys; cows will be cows. *Child Development* 80(2):461–481.

Teicher A. 2020. *Social Mendelism*. Cambridge University Press.

Tompkins R, Vasquez K, Gerdin E, Dunham Y, Liberman Z. 2023. Expectations of intergroup empathy bias emerge by early childhood. *Journal of Experimental Psychology: General* 153(11):2700–2714.

U.S. National Science Foundation 2023. Diversity and STEM. https://www.nsf.gov/reports/statistics/diversity-stem-women-minorities-persons-disabilities-2023

de Valadés D. 1579. *Rhetorica Christiana*. Petrus Jacobus Petrutius.

Van de Werfhorst HG. 2024. Is meritocracy not so bad after all? *American Sociological Review* 89(6):1181–1213. https://doi.org/10.1177/00031224241292352

Wade N. 2015. *A troublesome inheritance*. Penguin.

Waggoner MR, Uller T. 2015. Epigenetic determinism in science and society. *New Genetics and Society* 34(2):177–195.

Wang MM, Cardarelli A, Brenner J, Leslie SJ, Rhodes M. 2025. Maladaptive but malleable. *Child Development* 96(2):865–880.

West K. 2025. *The science of racism*. Picador.

Whitaker ED. 2000. *Measuring mamma's milk*. University of Michigan Press.

Wilkinson R, Pickett K. 2009. *The spirit level*. Bloomsbury Press.

Williams B. 2009. Deadly inequalities. In Kusmer KL, Trotter JW, editors. *African American urban history since World War II*. University of Chicago Press. pp. 142–159.

Williams DR, Lawrence JA, Davis BA. 2019. Racism and health. *Annual Review of Public Health* 40(1):105–125.

Williams MJ, Eberhardt JL. 2008. Biological conceptions of race and the motivation to cross racial boundaries. *Journal of Personality and Social Psychology* 94(6):1033–1047.

# 2 Essentialism in science textbooks

## Race and human variation

As seen in the last chapter, genetic essentialism is pervasive in contemporary culture, easily amplified by information on human biology in informal media, and rooted in a long history of scientific racism and sexism. This chapter shifts the focus to the role of formal education in shaping how students perceive human differences. It examines simplified explanations in biology and psychology textbooks that encourage genetic essentialism, their impacts on beliefs and attitudes, and alternative, anti-essentialist educational models.

Students come to the classroom with widely held determinist-essentialist conceptions that make it harder for them to learn about genetics and evolution and more susceptible to the potential for formal education to enhance essentialist reasoning (see Chapter 1 and Duncan and Reiser 2007). Genetic determinism obscures the multifactorial interactions that shape phenotypes, while the conception that fixed essences pass intact across generations interferes with understanding of evolutionary processes and intraspecies variation. Instead of counteracting these ideas, instruction in biology and medicine widely presents race as a risk factor for disease and health disparities as reflective of essential differences, reinforcing students' belief that race is biological (see Lujan and DiCarlo 2024).

A comparison between essentialism and another common conception about genetics, design teleology, highlights the tenacity of genetic essentialism in the face of formal education. Design teleology improperly assigns goal orientation to the emergence of traits and thereby obstructs understanding of natural selection. Stern et al. (2023) gave secondary school students in four grade levels instruments that assessed explicit essentialist and teleological conceptions, implicit conceptions, or both.[1] Analysis of the responses revealed no relationship between essentialist and teleological conceptions, similarly to a previous study described in Chapter 1 (Stern and Kampourakis 2017). There also was no association between students' implicit and explicit conceptions, whether essentialist or teleological, indicating that to assess students' conceptions it is necessary to examine both

DOI: 10.4324/9781003639923-2

types. Finally, there was an age effect for teleological conceptions but none for essentialist ones: design teleology weakened with increasing grade level, whereas essentialism remained relatively constant. In other words, learning about biology diminishes conceptions of teleology but has little or no effect on essentialism. Consequently, essentialism needs to be refuted directly since it does not respond to teaching against design teleology.[2]

Methods for confronting genetic essentialism are described later in this chapter, after a review of mainstream biology education from the middle-school to introductory university level. Two principal themes dominate instructional resources: Mendelian inheritance, and the seemingly straightforward biochemical line from genes to proteins and consequently to phenotypes. These elements appear across textbooks used throughout the world, many of which are translations of English-language texts (Castéra et al. 2008; Donovan et al. 2024; Guevara et al. 2023; Jamieson and Radick 2017; Schmid et al. 2022).

Where incomplete dominance, gene-gene interactions, polygenicity, and gene-environment processes are included, they appear at the end of a course or textbook, after the classic components have been thoroughly covered (see Dougherty et al. 2011; McCartney 2023; Redfield 2012; Sparks et al. 2020).[3] Treating the complexities of genetic inheritance and phenotypic expression as side issues creates the impression that single-gene causation with a strong, clear effect on phenotype is typical instead of rare. Words such as "cause," "determine," "gene for," and "control" reinforce the apparent directionality from genes to traits and make traits appear fixed and permanent. The focus on monogenic trait expression in individuals diverts attention away from evolutionary processes that operate upon the vast amount of phenotypic variation (and associated genotypic variation) within populations. Examples involving different rates of monogenic diseases across racial groupings join a deterministic view of causation to socially constructed race categories and thereby implying the existence of biologically meaningful subcategories of the human species. Race essentialism may be further fueled by coverage of genetic variation between continental populations without discussion of the far greater amount of variation within these same populations, or the fact that gene frequencies vary over time and in a continuous manner across geographical zones. The following pages explore these tendencies, their effects on student knowledge and attitudes about social inequality, and the development of alternative instructional approaches.

## Genetic determinism

An idea of the knowledge considered most essential by authors of textbooks and online educational materials may be gleaned from the assessment questions offered to high school and university students to test their learning. These questions focus predominantly on the role of genes in shaping phenotypic traits, contrary to published educational standards which emphasize the importance of teaching about environmental processes as part of a multifactorial approach. To illustrate, Schmid et al. (2022) analyzed more than 1,000 genetics questions for undergraduate students from the end-of-chapter sections of a widely used genetics textbook (Klug et al. 2019), four freely available published concept assessments, and peer-reviewed online curriculum materials from *CourseSource*. The researchers counted the number of questions that asked about how traits are inherited and expressed; and, among those questions, how many addressed multifactorial concepts. Questions that concerned inheritance and causality amounted to 27% of the total and were coded according to whether they addressed single genes, multiple genes, environment, environment plus genes, or gene-by-environment.

Of the questions that concerned inheritance, 95% asked about single or multiple genes as opposed to gene-environment processes. Of those which dealt with relationships between genotypes and phenotypes, 65% concerned single or multiple genes.[4] Combining the inheritance and causality questions, the researchers found that only 9% asked about genes plus environment, while 0.7% concerned gene-by-environment mechanisms. The questions that included an environmental component were more often related to causality than inheritance, suggesting a need for more of the questions on inheritance to test knowledge of gene-environment processes. The fact that questions involving environmental factors were primarily found in the genetics textbook and one of the published concept assessments (PUGGS) indicates that more of these questions are especially needed in the other sources.[5]

In the US, Mendelian inheritance and molecular genetics are the foundation for all standards for genetics instruction (Donovan et al. 2024). Dougherty et al. (2011) report that that these standards rarely conform to the American Society of Human Genetics' core concepts. After reviewing educational standards for high school genetics in all US states plus the District of Columbia, the authors conclude that 85% are inadequate, mainly because they do not go beyond Mendelian and molecular genetics to incorporate knowledge about multifactorial processes. Statewide assessment examinations to test student knowledge consequently channel genetics instruction toward the traditional approach.

International comparisons indicate that deterministic views of genetic causation are embedded in textbooks throughout the world. Castéra et al. (2008) analyzed 50 textbooks (1–11 per country) from 16 countries (12 European countries plus Lebanon, Senegal, Tunisia, and Morocco).[6] They found that books in all countries reinforced deterministic thinking through photographs of monozygotic twins uniformly pictured in the same body position and wearing the same clothes and hairstyles, conveying the idea that genes cause behavioral traits. However, the authors documented differences in the use of terms such as "genetic program" and "genetic instructions," which appeared consistently in books from Finland, France, Lebanon, and Morocco, but were nearly absent from Eastern European books (Estonia, Lithuania, Poland, Hungary).

The researchers explain that the lesser use of deterministic terms in Eastern European books may be attributable to the lasting impact of the mid-20th-century anti-evolutionary-science political movement led by the Soviet agrobiologist, Trofim Lysenko, which shut down genetics research and subjected thousands of mainstream scientists to dismissal, imprisonment, and, in some cases, execution. Deterministic terms were also infrequent in books from Germany, reflecting an enduring reaction to Nazi science, with no use of the expression "genetic program." Italian books also avoided the term, possibly for the same reason, but the Italian books included deterministic words such as "project" and "patrimony." In the remainder of cases, deterministic language varied but was nonetheless present (Malta, Cyprus, Portugal, Romania, Senegal, Tunisia). For instance, the books in Cyprus used "genetic information" but avoided "genetic program"; the Romanian book used both terms but did not develop the topic of human genetics; and, the Tunisian textbooks avoided the term "genetic program" and incorporated advanced knowledge more than French books but nevertheless expressed deterministic thinking in other ways.

To illustrate how textbooks incorporate misleading deterministic language, Donovan et al. (2024:824) provide an excerpt, followed by an alternative version. The first says,

> You might be surprised to learn that the ability to roll your tongue is an inherited trait. Being able to roll the tongue is likely to be dominant over not being able to roll it. Characteristics such as tongue rolling are called discrete, because they tend to fall into distinct, or discrete, categories. Either you can roll your tongue or you cannot. Either you are male or female. Either you have dimples or you do not.

The replacement version expresses scientifically accurate concepts, as follows:

> Characteristics such as tongue rolling vary continuously. For example, some individuals can roll their tongues more than others. Likewise, human sexual development is also somewhat continuous. For example, while genetic males and females are the most common sexes, scientists estimate that at least 2 out of every 10,000 humans are intersex.[7]

This example illustrates how textbooks can communicate more correct information while also discouraging deterministic reasoning through language that avoids either/or constructions.

## Essentialism

In addition to projecting a genetic-determinist view of causation, textbooks consistently illustrate genetic concepts through examples which suggest that racial groupings have a biological basis. An analysis of 80 high school biology textbooks published from 1952 to 2002 found that at the beginning of the 50-year period the books did not refer to race in discussions of genetic disorders, but by the end nearly all did. About 90% of those published from 1993 to 2002 included discussions of race (Morning 2008). The racial examples were selected strategically from the scientific literature and, in some cases, had no scientific basis. In short, by the turn of the 21st century, biology textbooks had firmly associated race with genetics.

An analysis of racial content in 11 US high school biology textbooks published from 2014 to 2019 found that only one of the texts openly discussed the topic of human races (Willinsky 2020). Nine of the 11 referred to race either directly or indirectly through explanations of diseases, evolution, skin color, or eugenics. As seen in the last chapter, a "backdoor vignette" about disease genetics is sufficient to generate increased belief in racial differences in cognitive and behavioral traits. Experiments with middle and high school students concur that exposure to basic genetics principles strengthens deterministic thinking and reinforces popular misconceptions about genetic inheritance, expression, and variation. Learning genetics through the Mendelian-molecular paradigm makes students more likely to extend essentialist beliefs about physical traits to mental and behavioral ones; more likely to attribute racial disparities to genes than environmental (physical, social, cultural) factors; more likely to perceive a greater amount of genetic difference between than within races; and, less likely to agree with measures to address racial disparities in education (Donovan et al. 2021).

Donovan (2014) demonstrated this effect through an experiment in which eighth-grade students were given one of two 30-minute text-based lessons on variation in rates of genetic disease across populations.[8] Both texts discussed sickle cell disease and cystic fibrosis, but only one included the racial labels, African American and Caucasian, respectively. Students in the racialized condition expressed stronger essentialist beliefs than those who read the version that was free of racial language. The students in the racialized condition showed stronger agreement both with belief in genetic causes of racial differences and the idea of races as fixed and distinguishable entities. Testing a week later showed that the effect persisted after the experiment ended.

A subsequent study using the same two texts yielded similar results: an increase in race essentialism, with a greater tendency to attribute disparities in achievement to genes. After completing an instrument assessing preexisting views on race, ninth-grade students read a text on Mendelian genetics, one with racial terminology, the other without (Donovan 2016). Upon finishing the reading, students answered two sets of questions: one designed to assess race essentialism; the other designed to probe how students interpret racial disparities in educational achievement and how personally invested they are in working toward closing the gaps. The students who read the text containing racial terminology expressed much stronger belief in both bio-somatic and bio-behavioral essentialism than the students presented the alternative text. In written explanations, only the students in the racialized condition mentioned genes in relation to racial achievement gaps. Around two-thirds of students in the racialized condition and three-quarters in the non-racialized condition offered social explanations related to values, culture, and parenting, or, less frequently, historical and present discrimination.

The experiment also found that students' willingness to devote time to addressing achievement gaps varied depending on their expressed agreement with either race constructionism or genetic essentialism at the beginning of the intervention (Donovan 2016). Among those who expressed essentialist views, the students who read the text with racial references showed less readiness to address achievement gaps than those in the nonracial condition. Among the students with a social constructionism vision of race, willingness was higher in the racialized condition than the other group. Donovan explains that this is likely due to the looser boundaries between groups in a social constructionist understanding of race, which makes it easier to extend favoritism to groups outside one's own. That is, among students who start out with a social constructionist view of race, activation of race essentialism by the racial-condition reading causes more willingness to fix achievement gaps than is prompted by the neutral presentation in the non-racial-condition reading.

In a subsequent study, students in grades 7–9 were taught four lessons that covered racial differences in disease frequencies and skeletal structure, using one of two texts (Donovan 2017). The first was a state-approved textbook; the second, a modified version of the same text in which racial terminology was removed or replaced by other words. The altered text eliminated the words "race," "Caucasians," "Ashkenazi Jews," and "African Americans" and inserted terms such as "Americans" and "healthy people" in their place. Compared to the students who read the altered text, those presented the original four lessons came to perceive far more genetic

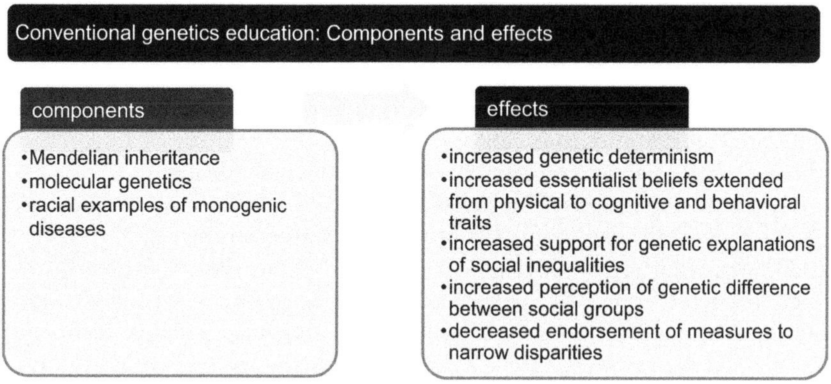


*Figure 2.1* Conventional genetics education.

distance between races by the end of the three months during which they were taught the target subjects. Their perception that genes determine racial differences in intelligence grew stronger. Finally, support for measures to mitigate disparities in education grew less among students who read the original text, as did the inclination to develop friendships with people of other races.

The above studies demonstrate that basic genetics education reinforces genetic determinism and essentialism and may encourage students to adopt these ways of thinking where they were previously undeveloped. The practice of illustrating genetic inheritance through racialized examples related to diseases and skeletal measurements gives students the impression that races are biologically meaningful groups of human beings. It allows them to conclude that if such groups differ in the conformation of their bones or their susceptibility to genetically determined diseases, then they must also differ in terms of behavior and cognition. As a result, students exposed to this kind of instruction are more likely to endorse eugenic beliefs, less likely to attribute racial disparities to historical and socioeconomic factors, less motivated toward cross-racial friendships, and less capable of learning about genetics and evolutionary biology (see Donovan 2022; Donovan et al. 2024; Stern et al. 2023). Figure 2.1 summarizes these effects.

## Basic genetics education overview

The following section reviews the components of basic genetics instruction to provide a baseline for the discussion in this and subsequent chapters. The two main components have remained stable for decades and are widely taught across countries starting at the middle school level. One is the laws of heredity described by the Augustinian friar, Gregor Mendel, in relation to experiments conducted at his Moravian abbey. Mendel's findings date to the 1850s and 1860s but did not spread through the scientific community until the turn of the 20th century. The other is the biochemical steps from genes to proteins, starting with the double-helical structure of the DNA molecule elucidated in the mid-20th century by an international set of researchers. The historical context for these discoveries is discussed below.

Mendel's experiments with pea plants aimed to create homozygous new plants from hybrids (heterozygotes) (see Kampourakis 2023; Radick 2023). Studying the proportions of seeds and plants with a variety of distinct characteristics that emerged from various crosses – wrinkled versus smooth seed case, yellow or green seeds, red or violet-red flower color, and so on – led him to surmise the existence of dominant and recessive "factors" that pass from one generation to the next. In addition, Mendel realized that the factors for a given trait segregate during the formation of gametes such that each gamete carries only one type, and that factors are inherited separately from one another through independent assortment.

Students are taught how to model the transmission of these factors, now known as alternative alleles of a given gene, with a chart invented early in the 20th century by the English geneticist Reginald Punnett and published with William Bateson in 1910. Along the top of the Punnett Square, students list the two alleles inherited from one parent (or pea plant), using a capital letter for a dominant allele, lower case for recessive. Along the left side, they do the same with the alleles inherited from the other parent. Then they pair the alleles in the four inner cells of the square. For instance, if an *AA* homozygote is crossed with an *aa* homozygote (as in Mendel's experiments), all offspring will be *Aa* and the recessive trait will not appear in the first generation. If two of these heterozygotes are then crossed (each carrying a dominant and recessive gene), then there is a one-in-four chance that the offspring will end up with two recessive alleles, a one-in-four chance that they will carry two dominant alleles, and a two-in-four chance that they will be heterozygous for the trait, like their parents. That is, the trait may reappear in the second generation, with a probability of one in four.

Next, a Punnett Square is used to model the transmission of two sets of alleles in a cross between two organisms that are heterozygous for both traits. This exercise illustrates the law of independent assortment. Along the top are the four possibilities from the first cross: *AA*, *Aa*, *Aa*, *aa*. Along the left side, *BB*, *Bb*, *Bb*, *bb*. Filling in the 16 cells yields a 1-in-16 probability of inheriting recessive alleles for both traits (*aa* and *bb*). There is a 9-in-16 probability of inheriting genotypes for neither recessive trait (*AABB*, *AABb*, *AaBB*, *AaBb*); a 3-in-16 probability of inheriting the recessive B trait without the A (*AAbb*, *Aabb*); and a 3-in-16 chance of inheriting the recessive A trait without the B (*aaBB*, *aaBb*).

Independent assortment generally results in the non-concordance of genes and their associated traits (for example, blood type and eye color). However, some alleles co-occur with relatively high frequency within families, such as alleles related to pale eyes and red hair (linkage). At the population level, in contrast, alleles at non-adjacent loci are not likely to be inherited together. Linkage disequilibrium expresses the amount of deviation from this expected non-association; that is, the strength of nonrandom associations in the pool of genomes making up a population.

Textbooks regularly provide human examples of traits presented as following a clear dominant/recessive pattern: detached or attached earlobes, dark or light hair color, a tongue that can roll or not. In addition to alleles on humankind's 22 pair of numbered autosomal (*auto*/same + *soma*/body) chromosomes, textbooks offer a similarly either/or description of sex-linked diseases and conditions related to genes on the 1 pair of allosomes (*allo*/different + *soma*/body), which are labeled with letters (see Chapters 4 and 5).

Other illustrations focus on populations with relatively high frequencies of monogenic diseases. These include breast and ovarian cancer among Ashkenazi Jews; cystic fibrosis among European Americans (often referred to as "Caucasians"); sickle cell disease in some African populations and their descendants; and Tay-Sachs disease among Amish people, French Canadians in eastern Quebec, and Ashkenazi Jews. While discussing forensic analysis to identify human remains, some textbooks assert that race can be determined through skeletal measurements involving the skull, mandible, and other bones, without explaining the unreliability of the method and its basis in humanmade typological schemes and unrepresentative reference populations.[9]

To connect Mendelian inheritance to natural selection (the Modern Synthesis), textbooks provide examples of genetic changes in response to altered environmental conditions, such as the industrial pollution in English industrial cities that covered trees in soot and made it easier for birds to see lightly colored peppered moths. Moths that were darker in color thanks to one or more melanin-related mutations were better camouflaged against the trees and became more numerous in proportion to lighter moths. After clean air legislation went into effect, the proportions reverted to high numbers of pale moths.

Besides selection against harmful genes, students learn concepts such as genetic drift, which refers to random cumulative fluctuations in gene frequencies due to chance events affecting reproduction, mortality, and the merging of maternal and paternal chromosomes upon fertilization. Thanks to drift, small populations diverge genetically more rapidly than large ones. Students are also taught the Hardy-Weinberg principle, which holds that, in large populations, gene frequencies remain constant in the absence of immigration, mutation, and selective pressure.

Just as textbooks present genetic inheritance through simplified Mendelian concepts, they describe the relationship between genes and phenotypes by way of a narrow 20th-century vision of the DNA molecule and the biochemical steps from genes to proteins. Students learn about the four types of nucleotides (nitrogenous base plus a phosphate group and five-carbon sugar molecule) that make up strands of DNA: adenine, cytosine, guanine, thymine, with adenine pairing with thymine and cytosine with guanine on the inside of the strands that make up the double helix. Textbooks describe DNA transcription, whereby a messenger RNA template (with the same units except that uracil takes the place of thymine) forms opposite a coding segment of the DNA strand. Three-base combinations code for 1 of 20 amino acids. Enzymes and the energy molecule, ATP, activate the amino acids to join with transfer RNA and create aminoacyl-tRNA units that are "read" by ribosomes. During translation, the ribosome discards the tRNA part of the unit and links the amino acid to others to create a polypeptide. Polypeptides become proteins when they fold into characteristic shapes that allow them to perform various functions within the cell. Instructions flow in a single direction, for proteins cannot transfer information amongst themselves or back to nucleic acids (the Central Dogma).

In addition, textbooks review the processes by which cells divide, comparing mitotic divisions of somatic cells that result in diploid cells (from the Greek *diploos*, for double) with a full complement of paired chromosomes, to meiotic divisions of progenitor sex cells that create haploid gametes (Greek *haplós* for single or simple) with only one of each pair of chromosomes. Before mitosis begins, the cell's chromosome pairs separate and are replicated to create two copies of each one, with DNA polymerase adding complementary base pairs and repairing errors where they occur. During mitosis, each of the two sets of pairs separates and the individual chromosomes are pulled apart by spindle fibers. At opposite ends of the cell, homologous chromosomes come into alignment and then the cell divides into two cells that each contain a nucleus with identical chromosomes and equal amounts of cytoplasm and organelles.

Meiosis in sex cells is different in several ways. The resulting cells contain only one of each chromosome, the chromosomes are not identical to the ones from which they were copied, and four cells are produced – although in females three of them do not develop into an ovum. During the pre-meiotic replication phase, crossing-over occurs between the chromosomes inherited maternally and paternally, creating novel combinations of DNA before each chromosome is replicated to yield two identical chromatids. In the first phase of meiosis, the replicated chromatids separate and align with their homologues; they undergo DNA cutting and recombination, which brings exchanges of DNA between them. This is followed by the phase in which the paired chromatids separate and segregate into four groups that become enveloped in nuclei before the cell divides into four haploid cells. Consequently, gametes contain an assortment of maternal and paternal chromosomes that have been altered by crossing-over and recombination. This shuffling creates genetic diversity across siblings and generations, resulting in variable phenotypic traits.

Textbooks also explain that the DNA in mitochondria – the energy-producing, formerly single-celled organisms engulfed by larger cells a billion years ago – passes only from mother to child. Egg cells contain hundreds of thousands of mtDNA molecules, whereas only a handful, if any, of the sperm cell's mitochondria enter the egg cell during fertilization. The latter carry mtDNA molecules that have degraded from the time spent in the male reproductive tract to their arrival in the ovum. As seen in the next chapter, given that is it inherited maternally without recombination with other DNA, mtDNA is useful for tracing genetic relationships, just as the Y chromosome sheds light on genetic relationships among males.

Finally, students learn that new alleles arise through mutations, which represent permanent changes to an organism's DNA (or RNA in the case of some viruses). These alterations may occur spontaneously or due to environmental exposures such as ionizing radiation or certain chemicals, and they may arise in both germ and somatic cells. Where mutations in the coding portion of DNA are not repaired, they alter the amino acid chain and thereby the associated protein. Mutations with catastrophic effects tend to disappear because the organism is not likely to reach reproductive age, which explains why serious genetic diseases are rare. Alternatively, mutations may confer an advantage and spread through a population over successive generations.

In addition to visual representations (drawings, electron micrographs, photographs) of DNA, organelles, cell division, and protein synthesis, textbooks include graphics such as illustrations of sickled versus normal red blood cells that concretize the racial characterization of disease. Pictures of physicians and patients or couples in consultation about genetic diseases substantiate a genetic-determinist view of causation. The sex/gender binary discussed in later chapters finds visual expression in pedigree charts that depict genetic inheritance with squares and circles for human males and females, but other symbols for nonhuman species, such as seeds in reference to Mendel's pea plant experiments.

Beyond principles of Mendelian and molecular genetics covered by basic genetics education and summarized above, genetics standards and curricular materials include teaching about multifactorial and dynamic inputs to phenotypic traits, especially at higher grade levels. However, the minimal and tacked-on coverage of these topics contributes to a presentation that simplifies complex biological systems, reinforces genetic determinism and essentialism, and renders multifactorial interactions sidebars instead of the core narrative. Basic genetics education thereby lends support to popular beliefs about genetically distinct races of humans. It also provides a foundation for attributing sex/gender differences to chromosomes and hormones (discussed in Chapter 4).

### Psychology: Mendelian inheritance meets another 19th-century genetics paradigm

While basic genetics education allows students to infer that physical differences between purported racial groups are indicative of cognitive and behavioral differences as well, psychology education provides an explicit bridge between the two in the form of human behavioral genetics. Its reach is vast, given the popularity of psychology courses among high school and university students. In the US, of the advanced-placement science exams secondary school students may take to skip first-level university courses, the psychology test is the one most frequently chosen. Nearly one-third of high school students take courses in psychology; psychology ranks among the top six undergraduate degrees awarded; and, college students from many other disciplines take psychology courses, including students who need them to satisfy a requirement for a degree in a health-related field (McCartney 2023).

By taking a Mendelian approach to inheritance that emphasizes dominant and recessive genes, single-gene disorders, and chromosomal anomalies, psychology textbooks present a gene-centric view of causation whereby development is a predictable process of unfolding genetic effects that move a single direction to stable, immutable traits. The "nurture" component is secondary and limited to an influence on the degree to which genes are expressed. Quantifying the additive contributions of each to human behavioral, personality, and cognitive traits is the task of human behavioral genetics, a discipline born in the late 19th century during an era of vivid enthusiasm for genetic determinism. Alcoholism, prostitution, criminality, and poverty were considered to derive from biological defects, which could now be characterized as genetic thanks to Mendel's discoveries (see Teicher 2020). One of the discipline's originators, Charles Darwin's younger half-cousin and inventor of the word "eugenics," Francis Galton, sought to demonstrate mathematically that genius and greatness were inborn traits, as if these characteristics arose independently of the inherited privileges and material advantages that he and other elite men enjoyed (see Fancher 2009).

Both the name of the discipline and its statistical measure are misleading. Human behavioral genetics is concerned with statistical relationships, not laboratory work on genes and the biochemical processes involved in the generation of a phenotype. Heritability ($H^2$) expresses a relationship between two measures of variance calculated for a specific population (see below). It does not describe inheritability or causation, the measures of which are penetrance (the proportion of individuals who carry both a gene and an associated trait) and expressivity (the degree to which a gene is manifested in a phenotypic trait), respectively. Heritability contrasts a genetic against an environmental component of a phenotypic trait, erroneously implying a fixed amount of genetic causation and reinforcing the timeworn nature/nurture dichotomy.

Heritability estimation was invented in the 1930s in agricultural studies to determine the degree to which selective breeding influenced a trait of interest. $H^2$ expresses how the variation in the genes of a sample of plants or animals and the variation in their environmental circumstances are mathematically related to variation in one of their phenotypic traits.[10] Heritability scores range from zero to one (or 100%) and are equal to genetic variance divided by phenotypic variance.[11] Subtracting this score from one yields the proportion of variance attributed to environmental factors, with the rest assigned to genes.[12]

Importantly, heritability statistics refer to the sample on which they are based and are not generalizable to other samples. They do not explain how genetic and environmental factors influence phenotypes or to what degree they matter. Estimates calculated for a particular trait may be very different from one study to the next and may change in the same sample between one point in time and another. The same heritability score can emerge from very different mechanisms, while cultural phenomena can either diminish or expand the genetic variance portion of heritability.[13] This means that policies such as educational interventions cannot be evaluated using heritability estimates, as if they were equivalent to artificial breeding in agriculture.

Given that genes and environments play a role in all traits, it is theoretically unsound and methodologically impossible to tease them apart. To illustrate, Moore and Shenk (2017) provide two illustrations. The first compares two children filling a bucket of water from different hoses, each with a measurable contribution, to the correct model for trait development in which one child holds the hose while the other turns the spigot. The second illustration, from Richard Lewontin, explains why heritability estimates cannot speak to the question of causation. First, an experimenter grows a batch of plants from genetically variable seeds in identical conditions suitable for growth (water, light, nutrients). Next, the experimenter grows a batch of plants from genetically variable seeds in identical but inadequate conditions. In both cases, the plants grow to variable heights. Given that the seeds were genetically identical, all the variability across the two cases is attributed to genetic variability, yielding a heritability score of one, or 100%. Yet, the plants cultivated in unfavorable conditions grew less, indicating that size is not due entirely to genes.

Many human examples demonstrate the pitfalls of using heritability estimates to analyze phenotypic variation. The monogenic disease, phenylketonuria (PKU), can have a heritability of zero where a population prevents the disease through screening and dietary modifications (see Chapter 3). Without these measures, heritability is one (100%) given that consumption of foods containing phenylalanine reliably results in neurological damage in affected individuals. The heritability of schizophrenia (up to 0.8 or 80%) is lower where genetically predisposed people are protected from the disease-driving stress of poverty by social systems and medical treatment that allow them to perform well at school and work. A comparison of Danish monozygotic and dizygotic twin pairs born between 1870 and 1900 yields a heritability score for longevity of 0.23 (23%), indicating a large environmental component (77%) that would be even greater had the sample been composed of unrelated people with fewer shared environmental factors, or had been collected during an era in which Danes were exposed to more severe health risks; among New York City residents in the 20th century, environment accounted for more of the variation in life span among socially subordinated groups such as African Americans (90%) than European Americans (70%) (Graves 2015).[14]

Even though heritability estimates are context-dependent and do not explain causation, they are discussed in the media and psychology texts as if they expressed the degree to which inherited genes determine a trait or behavior (see McCartney 2023). These statistics create the impression that, across all humans, the causes of complex, polygenic behavioral/cognitive traits such as intelligence, political orientation, aggression, or criminal behavior fall into discrete amounts split between genes and environment.[15] The first of these traits, intelligence, is of paramount concern to human behavioral genetics, given the positive association between scores on intelligence tests and measures of educational attainment, health, and occupational status and income. Analyses of average scores on intelligence tests feature prominently in academic and popular debates about the causes of racial disparities in opportunity and wealth (see Chapter 3).

However, current statistical methods are unable to elucidate the interplay between genetic and environmental factors in intelligence. Heritability estimates overemphasize the genetic contribution and efface the role of inherited environments (see Sauce and Matzel 2018).[16] Genome-wide association studies (GWASs) on unrelated individuals understate the heritability of intelligence scores, for scores increase where data are segregated by subpopulation, reflecting the impact of group-wide inherited environments. Contrariwise, twin and family studies tend to overestimate the heritability of intelligence scores due to the extra-genetic familial transmission of socioeconomic, cultural, and ecological conditions.[17] Parents also pass on non-genetic biological factors that affect development such as inherited epigenetic marks from both mother and father, and nutrients, microbes, and immunities from mother to offspring.

In addition, the assumption that intelligence can be accurately defined and measured is untenable for any one cultural or historical context and even shakier across contexts (see Sternberg 2007). Intelligence is highly dependent on individual characteristics and experience, mutable over the lifetime, and responsive to training as well as perceptions about its immutability or changeability. Cultural contexts define tasks and solutions that shape how intelligence develops in different societies or subpopulations within societies, as well as the kinds of intelligence that are valued, such as interpersonal ability, self-knowledge, emotional awareness, and problem-solving skills. Moreover, scores on intelligence tests vary with the closeness of fit between test designers and takers, as Ashley Montagu famously showed through a re-analysis of 1.75 million IQ scores from First World War military recruits in the US. At the time, the results were interpreted as proving the lower intelligence of the source populations of immigrants, due to a correlation between scores and the length of US residence of the men and their forebears. The re-analysis showed that White men from the North – the same category as psychologist Robert Yerkes, who designed the test – scored highest, followed by Northern Blacks, Southern Whites, and Southern Blacks (in Gould 1981). This effect contributes to historical shifts in national scores on intelligence tests (the "Flynn Effect") as well as cross-national variation, both of which are linked to a set of interrelated factors: years of schooling, literacy rates, and proportion of the population engaged in agriculture (Barber 2005).

Instead of confronting the complexity of intelligence and its measurement, psychology textbooks overwhelmingly fail to adequately define intelligence, discuss the limitations of intelligence testing, and present up-to-date models of intelligence. Of 29 top-ranked introductory psychology textbooks on Amazon.com, Warne et al. (2018) found that only 7 presented contemporary models of intelligence such as Charles Spearman's *g* factor.[18] This "general ability" score expresses the degree of agreement across an individual's performance on multiple, unrelated cognitive tasks and is considered highly predictive of academic ability. The remaining 22 books presented outmoded and/or nonmainstream theories. Of the total, 23 books contained logical fallacies and 23 included inaccurate statements.

The standard design for studies in behavioral genetics involves comparing a trait in people who share genes and environments in various proportions. In the classic model, environments are defined as 100% shared if children grow up in the same household and 0% shared if they grow up in different households. Shared gene fractions derive from the mechanics of sexual reproduction discussed earlier: one (100%), in the case of monozygotic twins; one-half among dizygotic twins and other siblings, and between parents and children; one-fourth among half-siblings; and so on. The frequency of a phenotypic trait is compared between sets of monozygotic and dizygotic twins, under the assumption that the environment plays an equal role in the expression of the trait in both cases and that the amount of contradictory or fluctuating influences on the trait is minimal. Alternatively, monozygotic twins raised in different households may be compared under the assumption that their shared genes express themselves in entirely non-shared environments.

Neither assumption about a child's environment is scientifically justified, nor are several others described below (McCartney 2023). Regarding the equal environment assumption, no environment remains constant over time; siblings alter family dynamics; and parents and other caregivers do not treat children identically (such as girls versus boys, later- versus earlier-born children, and more agreeable and conscientious versus harder-to-parent children).[19] Moreover, the prenatal environment in which twins develop is often different for one than the other. Like all dizygotic twins, one-third of monozygotic twins have separate placentas and chorion sacs, which exposes them to different hormone and growth factor levels. Contrariwise, the other two-thirds share a host of developmental factors in utero, which means that monozygotic twins raised apart cannot be said to share no aspect of their environment. Dizygotic twins raised in the same household do not share the same prenatal environment, nor do they share the same family environment, just like siblings who are not twins.

Similarly, the assumption that adoptive parents provide all environmental influences is untenable. Adopted children shared their prenatal environment with their biological mothers, with lasting effects on the children's physical and psychological development. Human behavioral genetics instead attributes any concordances between adopted children and their biological mothers to genes only. In addition, many other shared environmental influences are artificially excluded by the equally dissimilar environments assumption, including ubiquitous atmospheric/climatic conditions and cultural beliefs and practices, all of which impact the development and permanence of traits. In short, children who live in different households nonetheless share environments and experiences.

The complications listed above make it impossible for behavioral genetics studies to replicate the randomized, controlled trial design they purport to follow and cast doubt on any data they generate (McCartney 2023). The analogous procedure would involve defining different fractions of genes as the "treatment" condition equivalent to a new drug or procedure, with an entirely identical or completely dissimilar environment construed as the "control" condition, like standard care or a placebo. Moreover, studies involving adopted children cannot uphold the randomization requirement, given that adoptive parents represent a narrow range of economic, educational, social class, and other characteristics. They must meet strict criteria for consideration, are often matched with children based on race/ethnicity, and may be relatives of the adopted child.

The idea of an entirely shared environment in behavioral genetics leads to the conclusion that differences between racial groups living in the same country are due to genes. Alternatively, genes appear to be the exclusive reason for shared behavioral traits between monozygotic twins who grow up in different households or for differences between adoptive and biological children in the same household. In sum, together with Mendelian genetics, discussions of human behavioral genetics prioritize genetic influences while relegating environments to inconvenient clutter. Psychology textbooks thereby reinforce the essentialist misperception that genes determine traits in a permanent, invariable manner.[20]

Developmental systems involve multiple interlinked environmental, genetic, cellular, and behavioral factors that vary over time. Partitioning these influences into a genetic and environmental component is contrary to biological principles and methodologically unfeasible. A more processual understanding is needed and discussed further below.

## Anti-determinist, anti-racist genetics education

Besides reinforcing genetic essentialism in the ways discussed above, current biology textbooks from around the world provide contradictory and incomplete information that hampers learning. Many textbooks explain genetics concepts and models inconsistently and fail to explain the epistemic conditions under which they were discovered and developed (Dos Santos et al. 2012; Gericke et al. 2014). As a result, textbooks tend not to effectively integrate Mendelian and molecular genetics, leaving students with little understanding of the way proteins are synthesized and how they work at the cellular and organismal levels (see Donovan 2022; Guevara et al. 2023). Classroom instruction consequently may bring only limited improvement in student knowledge of genetics, as shown by a study in which a 31-item assessment instrument with basic questions about DNA, chromosomes, and proteins was given to hundreds of university students before and after six sections of a non-major course that included genetics content. The proportion of correctly answered questions rose only modestly, from 43% to 49% (Bowling et al. 2008).

As part of an analysis of group interviews with upper-level high school students to probe how they connect molecular genetics to protein synthesis, Gericke and Wahlberg (2013) mapped concepts into clusters united by central concepts such as DNA, gene, and protein. The analysis confirmed that students had trouble reasoning across clusters, compared to within them. Based on personal experience teaching university genetics courses, Redfield (2012) concludes that students exposed to a traditional university genetics curriculum learn what they must to complete exercises and answer test questions, yet their knowledge is constrained by the curriculum and soon forgotten. Students' scientific reasoning does not improve, and the course fails to satisfy their curiosity. Redfield finds that students want to know about the complex relationships linking genetic and non-genetic factors that affect most phenotypic traits. They seek genetic information that is relevant to their own lives and interests, such as ancestry, disease risk, criminal investigation, race, and behavior and cognition, and need guidance in interpreting media stories, scientific reports, and genetic tests. Donovan (2016) agrees that instructors need not avoid the topic of race so long as they also teach about the social construction of racial groupings and the biological realities that refute genetic essentialism.

The above discussion points to the importance of scientific literacy for absorbing new knowledge, especially information that conflicts with expectation. Indeed, the message that genetic essentialism is scientifically incorrect and socially damaging is often unconvincing to people whose knowledge base is unable to accommodate it. A refutational text that counters a simple view with one that is more nuanced may even cause the reader to end up linking myth with fact. Over time, the myth grows stronger (Donovan 2022).

Donovan (2022) distinguishes between two interrelated types of scientific literacy: a fundamental type that involves general comprehension and critical skills in reading scientific texts; and a derived type which concerns science knowledge. The first is necessary for developing the second, and the second makes it easier for the first to function with respect to assimilating new information in a text. Derived genetics/genomics literacy includes both ontological knowledge regarding genetic concepts and critical epistemological knowledge of the development and imperfections of scientific understanding. Basic biology education combines ontological knowledge of Mendelian inheritance, meiosis in sexual reproduction, and protein synthesis with epistemic knowledge of how these concepts arose through molecular genetics methods and dominant-recessive inheritance modeling. The combined effect is to present genes as a stable and immutable cause of traits, which channels causal reasoning toward genetic determinism. Groups appear to be genetically homogenous internally and genetically distinct from other groups, which favors essentialism. Armed only with this basic genetics literacy, a student who then learns that intelligence is heritable, and that races vary in scores on intelligence tests, may surmise that genes cause racial disparities.

The model known as "standard" genetics/genomics literacy, but which is much less widely taught than basic, includes ontological knowledge about phenotypic causation, as well as epistemological knowledge of quantitative genomics methods such as ancestry estimation and GWASs (Donovan 2022). Population thinking and polygenic approaches to trait variation highlight how genotypes and phenotypes vary in a continuous rather than dichotomous manner and consequently are inherently anti-essentialist. This kind of instruction tends to reduce deterministic and essentialist thinking, but modestly and not always. Where students lack the appropriate scientific literacy, learning about polygenic inheritance can end up bolstering genetic determinism and essentialism instead of appreciation of multifactorial processes.

A third approach, "humane" genetics/genomics literacy, is deliberately anti-determinist and anti-essentialist (Donovan 2022; Donovan et al. 2024). Humane genomics literacy includes all the ontological and epistemological content of basic and standard instruction, as well as identification and refutation of genetic essentialism. Race and racism are addressed directly because learning anti-essentialist concepts does not necessarily lead students to understand that, while race is socially constructed, it also has biological impacts. Moreover, learning that race is a social construct may contribute to belief in colorblindness instead of awareness that racial disparities stem from historically rooted sociopolitical structures and inequalities.

Humane genomics instruction refutes race essentialism by showing that nearly all the slight amount of variation between humans occurs within as opposed to between populations, and that frequency differences in common alleles account for much of the variation between populations. These facts invalidate the idea that most genetic variation occurs between races and that geographical populations construed as races are made up of individuals who are genetically homogeneous. Humane genomics instruction also demonstrates how complex traits depend on multiple genes and gene-environment processes, to reveal the absence of a scientific basis for attributing racial inequalities to genetic variation.

In a similar vein, McCartney (2023) advocates an "Epigenetics and developmental systems perspective" that focuses on dynamic and reversible epigenetic influences which shape genetic expression, as part of an interconnected system of genetic, environmental, and experiential factors. Using a "refutational method," teachers directly address students' misconceptions through a reading or activity that brings implicit beliefs to the surface and puts them in conflict with a scientific explanation that contradicts determinist/essentialist reasoning. For instance, instructors might counter misconceptions about the genetic determination of phenotypic traits through information on the effects of maternal starvation across multiple generations, or they might demonstrate how altered hormone levels during gestation may result in the development of traits not aligned with chromosomal sex.

Other authors suggest going beyond instruction on human genetic variation, gene-environment processes, epigenetics, and ancestry, to include a focus on the ways in which sociopolitical forces shape research questions, concepts, methods, and impacts (see Duncan et al. 2024). Historical examples include Linnaeus's classification of human races, with associated personality and behavioral traits, described in the last chapter, and Galton's class background and interest in proving the inheritance of qualities associated with elite status, mentioned above. The Mendelian model itself owes its flowering after a delay of decades to the emergence of public and scientific enthusiasm for eugenics in the early 20th century, as well as a personal rivalry between two men who had become friends at Cambridge: William Bateson, who allied himself with the Mendelian model; and the more successful evolutionary biologist, Walter Weldon, who propounded a theory of inheritance that emphasized gene-environment complexity in phenotypic causation. Stung by Weldon's criticism, Bateson redoubled his efforts to promote Mendelian heredity (Kampourakis 2023; Radick 2023; Teicher 2020). In the mid-20th century, sexism allowed men to take credit for Rosalind Franklin's work on the structure of DNA. James Watson and Francis Crick published their landmark paper without mentioning Rosalind Franklin; together with Franklin's colleague at King's College, Maurice Wilkins, they won a Nobel Prize. Among Franklin's previously unpublished work shared by Wilkins without her consent was X-ray diffraction photograph number 51, taken with graduate student Raymond Gosling (Franklin and Gosling 1953), which allowed Watson and Crick to infer the existence of base-pairing within the double-helical structure.

More recent examples connect the history of scientific racism to current manifestations and impacts. For instance, some of the twin studies that provide fodder for genetic-essentialist claims have been financed by racist organizations (Guevara et al. 2023; Marks 2017). The Texas Adoption Project and the University of Minnesota's Minnesota Study of Twins Reared Apart were supported by the Pioneer Fund, which was launched in 1937 with explicitly eugenic projects in mind, and which provided financial backing for the distribution of a Nazi propaganda film in the US. The Fund was headed by the outspoken scientific racist, J. Philippe Rushton (mentioned in Chapter 1), from 2002 to 2012. Although not motivated by racist ideology, the Human Genome Project was the fruit of a deterministic vision of genetic causation that engendered high expectations regarding the potential for DNA sequencing to definitively uncover everything possible about humankind. The outcome pointed to the extreme genetic similarity among all human beings and contradicted the biological race concept, yet genomic studies on racial differences did not disappear and instead continue to endorse the concept either knowingly or inadvertently (see Chapter 3).

Including a sociopolitical perspective in biology class would provide students the knowledge and evidence to critique the many manifestations of scientific racism, from historical eugenics arguments to contemporary claims about racial differences in intelligence and achievement (Guevara et al. 2023).[21] It would involve discussing the work of opposition scientists and anthropologists who have, since the beginning, refuted the claims of eugenicists and scientific racists. In addition, students would learn to critically examine the theory and practice of contemporary genomics research with regard to how populations are defined and by whom; which populations are chosen for study and how they are recruited; and, how environments are characterized, data are collected and analyzed, and benefits and potential harms are socially distributed (see Duncan et al. 2024; Sparks et al. 2020).

As part of teaching students that their fields are not culture-free, Sparks et al. (2020) urge science instructors to explore the large gap between the number of people from underrepresented communities in the overall population versus the scientific workforce noted in Chapters 1 and 5. This would involve investigating the historical and structural factors that have long favored White men's educational advancement and penalized other groups through differential access to resources and school systems funded by highly variable local tax bases. The authors encourage instructors to examine the relationship between their own career path and their personal identity and sociocultural context, and to address the ways in which their expectations of students and portrayal of science as value-neutral or culture-free may be an impediment to engaging students from underrepresented populations in science courses. The authors point out that "culturally relevant pedagogues" who work with rather than avoid differences of identity and experience are more effective instructors. Their acknowledgment of cultural factors in the history and current conduct of science welcomes rather than alienates students from underrepresented communities.

To broaden participation in science among people of all identities and backgrounds, Hales (2020) offers several suggestions for removing non-neutral ways of dealing with difference in relation to concepts in human biology. These include coverage of population genetics to counterbalance discussions of single-gene diseases through racial examples; inclusion of transgender individuals in pedigree charts; discussions of sex that go beyond binary X/Y classifications; and removal of stigmatizing language from explanations concerning genetic conditions and diseases, to validate rather than alienate people with visible and invisible disabilities.

In addition to improving students' ability to learn about human biology, as opposed to simply answering test questions correctly, the models discussed above foster critical awareness of the way in which genetic essentialism may be embedded in the conduct of science and the construction of knowledge about human variation. Showing how scientific fields can play a role in combating injustice and inequality has the further benefit of fostering an environment that favors retention of students from underrepresented communities in science majors and careers. Guevara et al. (2023) point out that these students are more inclined than students from dominant groups to pursue degrees in subjects that have a societal impact. As scientists, they not only expand the range of perspectives and research questions but also tend to be more aware of community interests with respect to research design, conduct, and impacts; attuned to ethical questions concerning data sovereignty; and effective in garnering and engaging community support for genomics research among historically marginalized populations.

Anti-essentialist genetics/genomics instruction improves not only students' immediate understanding of the material they learn in class, but also their long-term scientific literacy. It aligns genetics instruction with the information students learn in other classes, as opposed to contradicting it by passively or actively promoting genetic determinism and essentialism. As shown below, students respond to enhanced biology instruction with a reduction in race essentialism, beliefs in race colorblindness, and genetic-determinist reasoning.

## Effects of alternative models on perceptions of genetic determinism and essentialism

Providing students with more than the basic form of genetics education consistently results in lower agreement with genetic determinism and essentialism. Even brief exposure to material that explains human variation can correct students' misconceptions about race as a biological concept, as shown by a two-year study in which undergraduate students in multiple sections of an introductory biological anthropology courses were presented an explanation of four concepts related to human biological and genetic variation (Hubbard 2017). Students used clickers to respond to questions testing for content knowledge before and after the intervention. Questions included "Are there groups of (biological) traits that collectively can be used to define a racial group?" and "Are there more biological differences between racial groups or between individuals within a single race?" Comparing the answers before and after the intervention showed that understanding of common conceptual errors undergirding race essentialism increased significantly due to exposure to instruction on human biological variation.

To assess the effect of traditional versus more nuanced biology instruction on genetic determinism, Jamieson and Radick (2017) analyzed reactions to two versions of an introductory undergraduate genetics course.[22]

The two groups of students did not differ in gene-centric beliefs on a pretest at the beginning of the study. One of the classes was taught the conventional Mendelian model. The other class was modeled on the approach developed by Walter Weldon mentioned above. The Weldonian course used human examples to the extent possible, emphasized phenotypic variability and environmental contexts for genetic expression, stressed developmental processes from fertilized egg to multicellular organism, and avoided deterministic language. At the end of the study, there was no change in genetic determinism among students in the first class, but a significant decline among those in the second.[23]

A series of studies conducted with middle and secondary school students illustrates how teaching multifactorial genetics and population thinking changes the way people interpret within-group uniformity and between-group difference, reduces their agreement with genetic determinism and race essentialism, and facilitates learning by way of improved genetics literacy. The first of these involved teaching eighth and ninth grade students in two schools a week of lessons on variability in either climate (temperature and precipitation within and between zones) or human phenotypic and genotypic traits (within and between US census races) (Donovan et al. 2019). After the intervention, those in the first group scored much lower on assessments of perceived human genetic variation and demonstrated stronger cognitive biases related to race prejudice. The researchers then adapted the unit to a 45-minute computer-based lesson and obtained the same result through trials involving adults recruited through the Mechanical Turk website (Amazon) as well as students in 9th–12th grade from five schools in four US states.[24]

Further analysis of the data from the study with 9th–12th graders indicates that, in addition to showing greater declines in genetic essentialism upon learning about genetic variation within versus between populations, students with higher genomics literacy undergo steeper declines in perceptions of racial differences than students with lower genomics literacy (Donovan et al. 2021). The authors calculate that 1 of 40 students with low standard genomics literacy will come to reject genetic essentialism after a humane genomics intervention, whereas the ratio is only 1:12 among students with high standard genomics literacy. Students in the latter group are better equipped to engage with the curriculum materials and develop an understanding of the anti-essentialist implications of population genetics.

Using material modeled on that developed for the 2019 study, Donovan et al. (2020) engaged middle and high school students (grades 7–12) in six weeks of standard genetics/genomics education on multifactorial causation followed by humane genomics instruction regarding genetic variation within and between populations.[25] All ten classrooms started with basic genetics education. Of these, six continued with it while four learned the standard plus humane genomics curriculum. At the end of three months, the students in the treatment condition demonstrated greater improvement in multifactorial genetics knowledge and more significant declines in belief in genetic essentialism, relative to the students taught the basic curriculum. That is, students who had gained knowledge about multifactorial causation before being taught about within- versus between-group variation were more receptive to learning that there is little genetic variation between races.

The above studies did not assess whether social desirability bias influenced agreement or disagreement with essentialist beliefs and were not designed to evaluate whether reduced genetic essentialism leads to improved understanding of the social construction of race. These issues prompted Donovan et al. (2024) to demonstrate that exposure to humane genomics education results in lower adherence to genetic essentialist thinking, more openness to socializing with people of other races, and greater commitment to policies designed to address racial disparities. Classrooms of 7th–12th graders in multiple US states were divided between ones taught basic genetics education followed by humane genomics education and humane followed by basic education.[26] Fifteen teachers received training in humane genomics education for 40 hours. Their classes were divided into two groups: 25 were taught basic followed by humane genomics education, while 26 were taught humane followed by basic genetics education. The six-week period of instruction was divided equally between the two types of education.

In the classroom, teachers avoided using terms such as genetic essentialism, social constructionism, or colorblindness. To avoid directing students' perceptions away from genetic essentialism, they did not teach about the social construction of race. They did not discuss monogenic diseases in racial terms in either phase of the experiment, which allowed the study to verify that the reduction in genetic essentialism was due solely to humane genomics education rather than possible comparison effects. The researchers tested students three times: before the experiment; halfway through; and, after it was concluded. The questions elicited knowledge of genetic inheritance, genetic expression, and genetic/environmental processes; beliefs about genetic differences between races; adherence to genetic essentialism; and beliefs about social constructionism and race colorblindness. In addition, the researchers assessed the students' emotional responses to the material they were presented.

Comparing scores between the beginning of the trial and the end, both groups increased their knowledge of genomics and Mendelian genetics, belief in social constructionism, and tendency to make environmental attributions. Genetic attributions, perceptions of racial discreteness, and belief in genetic essentialism declined in both groups. Belief in race colorblindness remained relatively stable.

There were several differences between the two groups, such as a slight increase in belief in race colorblindness in the humane followed by basic group (hereafter H-to-B) and a decrease in the basic followed by humane group (B-to-H). The two groups finished at the same point with respect to Mendelian genetics knowledge, genomics knowledge, and racial discreteness, but arrived there by different routes. The H-to-B group's knowledge of Mendelian genetics declined to the half-way point, then rose quickly during the second half of the trial, whereas the B-to-H group's knowledge rose steadily over both phases. The H-to-B group's knowledge of genomics rose quickly in the first half and more slowly in the second, while the B-to-H group's knowledge rose slowly during the first half and quickly during the second. The H-to-B group's perception of racial discreteness fell rapidly during the first half and rose slightly during the second, whereas the B-to-H group's perception of racial discreteness fell slowly during the first half and rapidly in the second.

End-point scores on the other measures varied slightly. Genetic attributions fell quickly in the H-to-B group before rising during the second phase to a point above that of the B-to-H group, whereas genetic attributions in the B-to-H group fell little during the first phase and much during the second. Environmental attributions rose rapidly in the H-to-B group before declining to a point below that of the B-to-H group, whose environmental attributions rose slowly first and quickly second. For genetic essentialism, scores in the H-to-B group declined moderately in the first phase before inching up to a position above those of the B-to-H group, whereas the B-to-H group's adherence to genetic essentialism was steady in the first phase before falling in the second. Finally, belief in social constructionism rose quickly in the H-to-B group before declining slightly in the second phase to rest below that of the B-to-H group, whose belief in social constructionism declined slightly during the first phase before rising to a higher point in the second.

These results are not surprising, given that learning Mendelian genetics favors genetic determination and race essentialism, whether presented first or second. At the same time, they suggest that presenting Mendelian genetics first reduces the overall impact, compared to exposure to Mendelian genetics after humane genomics education. In both cases, exposure to the humane version benefits students with respect to learning about genetic inheritance and expression, notwithstanding the more complex concepts and information involved.

The increase in acceptance of social constructionism occurred without any leading on the part of teachers, indicating that it was the result of learning the principles conveyed by humane genomics education. Moreover, social desirability did not affect self-reported belief in genetic essentialism. The results did not vary between states, indicating that the findings may be generalized to many other states.

The researchers found no difference in students' experience of confusion, anxiety, or frustration between the two types of education. Students of color reported lower levels of negative emotion while learning humane genomics education than basic, whereas there was no difference in emotional experience between the two phases of the experiment among White students. This suggests that, for students of color, humane genomics education is especially favorable to learning.

The researchers followed up with another experiment that involved giving undergraduate students 60–minute online lessons covering concepts in humane genomics education (Donovan et al. 2024). While the study did not assess whether students who read the lessons developed improved knowledge of genetics and genomics relative to students not exposed to the lessons, it confirmed the other findings of the experiment described above. That is, humane genomics education brings a reduction in students' perception that races are genetically distinct; a shift in beliefs about the causation of complex traits from genes to combined genetic and environmental processes; and a reduction in agreement with genetic essentialism. Analysis of the effects of the various components of humane genomics education indicates that learning about multifactorial inheritance has a smaller dampening effect on genetic essentialism than reading about genetic variation between races. The study did not assess emotional responses or effects on colorblindness or social constructionism but found that students' cultural values and associated political beliefs did not alter the effect of the lessons on genetic essentialism. Some of the effects of anti-essentialist instruction are summarized in Figure 2.2.

Based on the above results, Donovan et al. (2024) conclude that the most effective way to teach genetics with an eye to reducing genetic essentialism is to first present the basic form, then discuss the limitations of that approach, and proceed with humane genomics education. Teaching multifactorial concepts and population thinking is beneficial in either order.

Other researchers advocate starting out with standard genetics/genomics education before basic, although these proposals contrast basic to standard rather than the humane genetics/genomics approach that includes everything in the standard model (see Schmid et al. 2022; McCartney 2023; Redfield 2012). Jamieson and Radick (2017) argue for avoiding a simple-to-complex curriculum because of the potentially enduring impact of the simple model on later exposure to more complex concepts. They recommend an inverted-curriculum approach that could be used beginning in middle school, with standard genetics education first, Mendelian genetics second. This would involve introducing students to complex concepts

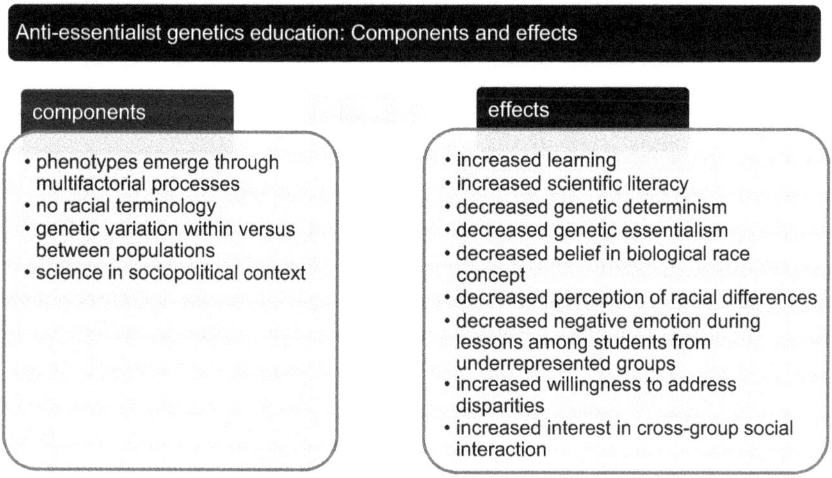


*Figure 2.2* Anti-essentialist genetics education.

at the beginning, through examples involving continuous polygenic traits such as height where the influence of environmental factors is evident, and by exploring the action of proteins within the cell to explain how genes relate to phenotypes. Next, courses would teach about monogenic traits that illustrate Mendelian principles and explain that even their expression depends on many factors. This method would better prepare students to understand genetics information they encounter outside the classroom and avoid strengthening the deterministic thinking by which traits are seen as either entirely genetically or completely environmentally determined (Dougherty 2009).

## Summing up

Science textbooks that offer simplified explanations of genetics and human variation advance a genetic essentialist approach to interpreting human differences that reinforces messages students receive from sources outside of school. Together these channels of information sustain the essentialist assumption that shared genes generate groupwide characteristics and account for perceived differences between groups. This chapter has shown that a more comprehensive genetics/genomics education can weaken belief in genetic determinism and race essentialism. Students exposed to humane genomics education are less likely to believe that complex traits derive from genes alone and more attuned to the interplay among multiple genetic and environmental influences. They are less likely to conclude that racial groups diverge genetically, less likely to agree with the denialism inherent in professed colorblindness, and more aware of the social construction of race. Humane genomics education satisfies students' curiosity about human biology, opposes the genetic essentialism that undergirds racist ideology and action, and equips students to make sense of genetics information they encounter outside the classroom.

Science textbooks tend to be slower to incorporate new discoveries into their explanations than textbooks in other fields (see Chapters 4 and 5). This conservatism holds formal science education back from challenging genetic essentialism. At the same time, the authority culturally associated with science lends science textbooks extra power to reinforce folk concepts about race. The next chapter develops an interdisciplinary, anti-essentialist counternarrative that challenges race essentialism and lays the groundwork for subsequent chapters, which show that simplified presentations of sex determination and human evolution reaffirm popular beliefs about inborn gender traits and their ultimate origins in prehistoric subsistence roles.

## Notes

1 Pooled data from schools in Switzerland and France. Unless otherwise noted, the studies cited in this chapter were conducted in the US.
2 As a step in the right direction, the authors suggest describing species with plural instead of singular nouns, to allude to internal variability and the possibility of evolutionary change.
3 Jamieson and Radick (2017) point out that the word "processes" is more accurate and less deterministic than "interactions," which implies that genes and environments are separate influences on development.
4 Sixteen percent of the questions concerned inheritance, while 11% concerned causality.
5 For differences in the distribution of various kinds of questions by source, see Schmid et al. (2022).
6 Date range 1983–2006, with the vast majority published between 2000 and 2006.
7 The proportion of people with chromosomal and hormonal variations from the norm is much higher than that of individuals strictly defined as intersex, as seen in Chapter 5.
8 San Francisco Bay Area. Unless otherwise noted, the studies by Donovan and Donovan and colleagues were conducted in private, public, or both types of middle and/or secondary schools in the San Francisco Bay Area.
9 For instance, the Fordisc series includes data on skeletal remains of 19th-century indigenous North Americans, skulls excavated from a nearly 2,000-year-old Egyptian cemetery, and the remains of male victims of the Khmer Rouge's "killing fields" massacre, classified as Vietnamese (see Clever and Jong 2025).
10 Moore and Shenk (2017) attribute the statistic to JL Lush. Visscher et al. (2008) credit Sewall Wright and Ronald Fisher.
11 Variance expresses the dispersal of values around the average. It is equal to the sum of the distance of each data point from the mean in both directions, squared, and divided by the number of data points.

12 Originally, genetic variance was a single measure that encompassed additive, dominance, and epistatic variance. Today, these components are quantified individually. On heritability calculations, see Barry et al. (2023); Turkheimer (2024).
13 For instance, changing norms in the mid-20th century US brought a rise in heritability estimates for fertility that unmasked genetic variation, whereas China's one-child policy resulted in a reduction in heritability that concealed genetic variation (Lala and Feldman 2024).
14 The Danish statistics are based on a correlation in life spans of 0.20 among monozygotic twins and 0.06 among dizygotic twins.
15 For instance, see Tuvblad and Beaver (2013) regarding aggressive antisocial behavior.
16 For example, see textbooks and other publications by Robert Plomin, such as Ashbury and Plomin (2013).
17 For instance, heritability of IQ drops from 0.5–0.8 to 0.3–0.4 when estimates account for familial cultural transmission (Lala and Feldman 2024).
18 Date range 2011–2017.
19 Each of the first in these pairs tends to be favored by parents, although where favoritism is based on autonomy parents favor earlier-born children (who tend to be given more independence), as shown by a meta-analysis of 30 studies and dissertations/theses involving around 20,000 participants (Jensen and Jorgensen-Wells 2025).
20 See Johnson et al. (2009), who argue that further heritability estimation of individual behaviors has limited value given that everything is heritable, although twin studies remain useful for identifying and assessing the weight of various constellations of environmental factors that impinge upon traits such as school performance and risk-taking behavior.
21 See Guevara et al. (2023) for four teaching modules, including worksheets and slide decks on: human evolution and biological variation; genotype-phenotype connections (using examples of human traits); ancestry; and scientific racism.
22 England (University of Leeds).
23 Donovan (2022) notes that, while the study has a few methodological shortcomings (such as the comparison between first-year biology students in the Mendelian course and second-year humanities students in the Weldonian course), it nonetheless supports the effectiveness of a complexity-first approach.
24 California, Colorado, Massachusetts, Washington DC.
25 Colorado and California (San Francisco Bay Area).
26 New Jersey, Kansas, Massachusetts, Illinois, Colorado, Indiana.

## References

Ashbury K, Plomin R. 2013. *G is for genes: The impact of genetics on education and achievement*. Wiley Blackwell.

Barber N. 2005. Educational and ecological correlates of IQ. *Intelligence* 33(3):273–284.

Barry CJ, Walker VM, Cheesman R, Davey Smith G, Morris TT, et al. 2023. How to estimate heritability. *International Journal of Epidemiology* 52(2):624–632.

Bowling BV, Huether CA, Wang L, Myers MF, Markle GC, et al. 2008. Genetic literacy of undergraduate non–science majors and the impact of introductory biology and genetics courses. *BioScience* 58(7):654–660.

Castéra J, Clément P, Abrougui M, Nisiforou O, Valanides N, et al. 2008. Genetic determinism in school textbooks. *Science Education International* 19(2):163–184.
Clever I, Jong L. 2025. Old bones in new databases. *American Anthropologist* 127(3):566–580.
Donovan BM. 2014. Playing with fire? *Journal of Research in Science Teaching* 51(4):462–496.
Donovan BM. 2016. Framing the genetics curriculum for social justice. *Science Education* 100(3):586–616.
Donovan BM. 2017. Learned inequality. *Journal of Research in Science Teaching* 54(3):379–411.
Donovan BM. 2022. Ending genetic essentialism through genetics education. *Human Genetics and Genomics Advances* 3:1–13.
Donovan BM, Semmens R, Keck P, Brimhall E, Busch KC, et al. 2019. Toward a more humane genetics education. *Science Education* 103(3):529–560.
Donovan BM, Weindling M, Amemiya J, Salazar B, Lee D, et al. 2024. Humane genomics education can reduce racism. *Science* 383(6685):818–822.
Donovan BM, Weindling M, Lee DM. 2020. From basic to humane genomics literacy. *Science & Education* 29(6):1479–1511.
Donovan BM, Weindling M, Salazar B, Duncan A, Stuhlsatz M, et al. 2021. Genomics literacy matters. *Journal of Research in Science Teaching* 58(4):520–550.
Dos Santos VC, Joaquim LM, El-Hani CN. 2012. Hybrid deterministic views about genes in biology textbooks. *Science & Education* 21:543–578.
Dougherty MJ. 2009. Closing the gap. *The American Journal of Human Genetics* 85(1):6–12.
Dougherty MJ, Pleasants C, Solow L, Wong A, Zhang H. 2011. A comprehensive analysis of high school genetics standards. *CBE—Life Sciences Education* 10(3):318–327.
Duncan RG, Krishnamoorthy R, Harms U, Haskel-Ittah M, Kampourakis K, et al. 2024. The sociopolitical in human genetics education. *Science* 383(6685):826–828.
Duncan RG, Reiser BJ. 2007. Reasoning across ontologically distinct levels. *Journal of Research in Science Teaching* 44(7):938–959.
Fancher RE. 2009. Scientific cousins. *American Psychologist* 64(2):84–92.
Franklin RE, Gosling RG. 1953. Molecular configuration in sodium thymonucleate. *Nature* 171(4356):740–741.
Gericke N, Wahlberg S. 2013. Clusters of concepts in molecular genetics. *Journal of Biological Education* 47(2):73–83.
Gericke NM, Hagberg M, dos Santos VC, Joaquim LM, El-Hani CN. 2014. Conceptual variation or incoherence? *Science & Education* 23:381–416.
Gould SJ. 1981. *The mismeasure of man*. WW Norton & Company.
Graves Jr JL. 2015. Great is their sin. *The Annals of the American Academy of Political and Social Science* 661(1):24–50.
Guevara E, Gopalan S, Massey DJ, Adegboyega M, Zhou W, et al. 2023. Getting it right. *Biology Methods and Protocols* 8(1):bpad032.
Hales KG. 2020. Signaling inclusivity in undergraduate biology courses through deliberate framing of genetics topics relevant to gender identity, disability, and race. *CBE—Life Sciences Education* 19(2):1–9.
Hubbard AR. 2017. Testing common misconceptions about the nature of human racial variation. *The American Biology Teacher* 79(7):538–543.

Jamieson A, Radick G. 2017. Genetic determinism in the genetics curriculum. *Science & Education* 26(10):1261–1290.
Jensen AC, Jorgensen-Wells MA. 2025. Parents favor daughters. *Psychological Bulletin* 151(1):33–47.
Johnson W, Turkheimer E, Gottesman II, Bouchard Jr TJ. 2009. Beyond heritability. *Current Directions in Psychological Science* 18(4):217–220.
Kampourakis K. 2023. *How we get Mendel wrong, and why it matters*. CRC Press.
Klug WS, Cummings MR, Spencer CA, Palladino MA, Killan D. 2019. *Concepts of genetics*. Pearson.
Lala KN, Feldman MW. 2024. Genes, culture, and scientific racism. *Proceedings of the National Academy of Sciences* 121(48):e2322874121.
Lujan HL, DiCarlo SE. 2024. Misunderstanding of race as biology has deep negative biological and social consequences. *Experimental Physiology* 109(8):1240–1243.
Marks J. 2017. *Is science racist?* John Wiley & Sons.
McCartney J. 2023. Exposing the hazards of teaching 19th century genetic science. *Journal of Research in Science Teaching*. https://doi.org/10.1002/tea.21895
Moore DS, Shenk D. 2017. The heritability fallacy. *Wiley Interdisciplinary Reviews: Cognitive Science* 8(1–2):e1400.
Morning A. 2008. Reconstructing race in science and society. *American Journal of Sociology* 114(S1):S106–137.
Radick G. 2023. *Disputed inheritance*. University of Chicago Press.
Redfield RJ. 2012. "Why do we have to learn this stuff?"—a new genetics for 21st century students. *PLoS Biology* 10(7):e1001356.
Sauce B, Matzel LD. 2018. The paradox of intelligence. *Psychological Bulletin* 144(1):26–47.
Schmid KM, Lee D, Weindling M, Syed A, Agyemang SL, et al. 2022. Mendelian or multifactorial? *Journal of Microbiology & Biology Education* 23(3):e00093.
Sparks RA, Baldwin KE, Darner R. 2020. Using culturally relevant pedagogy to reconsider the genetics canon. *Journal of Microbiology & Biology Education* 21. https://doi.org/10.1128/jmbe.v21i1.1901
Stern F, Kampourakis K. 2017. Teaching for genetics literacy in the post-genomic era. *Studies in Science Education* 53(2):193–225.
Stern F, Kampourakis K, Müller A. 2023. "Genes for a role," "genes as essences." *Journal of Research in Science Teaching* 60(2):237–267.
Sternberg RJ. 2007. Who are the bright children? *Educational Researcher* 36(3):148–155.
Teicher A. 2020. *Social Mendelism*. Cambridge University Press.
Turkheimer E. 2024. *Understanding the nature–nurture debate*. Cambridge University Press.
Tuvblad C, Beaver KM. 2013. Genetic and environmental influences on antisocial behavior. *Journal of Criminal Justice* 41(5):273–276.
Visscher PM, Hill WG, Wray NR. 2008. Heritability in the genomics era—concepts and misconceptions. *Nature Reviews Genetics* 9(4):255–266.
Warne RT, Astle MC, Hill JC. 2018. What do undergraduates learn about human intelligence? *Archives of Scientific Psychology* 6(1):32–50.
Willinsky J. 2020. The confounding of race in high school biology textbooks, 2014-2019. *Science & Education* 29:1459–1476.

# 3 Human origins, genetic variation, genotype-phenotype relationships

The last chapter examined how simplified explanations of genetics favor essentialist reasoning, whereas more complete instruction results in reduced genetic essentialism and increased awareness of and willingness to address social, educational, and economic disparities. This chapter brings together several lines of evidence against essentialist ideas about racial differences in physical and behavioral traits: human origins and migrations; genetic variation within versus between populations; and the dependence of phenotypes on complex gene-environment-experience processes. It analyzes how genomic information can both shed light on disease processes and evolutionary history and be segregated in ways that recapitulate and reinforce a socially damaging view of difference.

## Human origins and migrations

The remarkable genetic uniformity of the human species reflects its recent origin and a lack of barriers to migration and mixing. The result is too little variation between groups to merit division into a small number of races. The following overview traces this history and its implications, which are relevant also to the discussion of gender and social inequality in Chapters 6 and 7.

Some time around 8–7 million years ago (mya), and possibly earlier, the human lineage diverged from chimpanzees. The ancestral species to the two had diverged millions of years previously from the other apes (gorillas, orangutans, gibbons, in descending order), which had separated from the Old World monkeys before 25 mya. Their relationships are evident in decreasing genetic similarity between humans and these respective species, all of which share features including powerful vision, reduced sense of smell, apposable thumbs, and facial musculature that favors social interaction and communication.

Before the human and chimp lineages split, there was a shared ancestral species that climbed trees, walked on two feet, and waded into rivers and

DOI: 10.4324/9781003639923-3

lakes, as do many nonhuman primates today. While common chimpanzees (*Pan troglodytes*) are widely regarded as a model for the last common ancestor with humans due to mixture of observations and suppositions that prioritize hierarchical, competitive behaviors and male dominance (see Chapter 7), the other, more cooperative chimp species, bonobos (*Pan paniscus*), is equally closely related to humans. Neither one lives in the kind of mosaic, highly variable environments inhabited by early human ancestors (Martin 2023). Human ancestors exploited innumerable habitats and developed myriad systems of food procurement, kinship, and social organization, whereas chimpanzees of both kinds inhabit a more restricted environment and remain more behaviorally consistent. In addition, not only are the remaining apes unrepresentative of the wider range of species that have existed over time, but current chimpanzee characteristics may have arisen after the split from the common ancestor.

The earliest human ancestors emerged during the period from 7 to 4.4 mya and include *Sahelanthropus tchadensis*, *Orrorin tugenensis*, and *Ardipithecus ramidus*. Like them, species of the *Australopithecus* (4.2 to 2 mya) and *Paranthropus* genera (3 to 1.8 mya) had traits favorable to both arboreal and terrestrial movement, and walked bipedally in a variety of ways and to varying degrees (Kirinya et al. 2024). The human lineage, *Homo*, was fully bipedal, with relatively long legs, short arms, and a spine, pelvis, and non-apposable big toe that supported upright posture as the primary form of locomotion. *Homo* first appeared around 2.8 mya in East Africa (see below). Evidence of *Homo erectus* (or *H. ergaster*, the same or a closely related species) from around 2 mya has been found in eastern, southern, and northern Africa, and, depending on the location, includes stone tools, cut marks on fossil bones, and skeletal remains. More than 1.7 mya, *H. erectus* left Africa to spread through Europe, the Middle East, and Southeast Asia (see Rogers et al. 2020).

The emergence of *Homo* corresponds in time with a period in which glacial expansion and the tectonic fracture of the Rift Valley in East Africa had reduced forest cover and brought more extensive grasslands. This did not confine early humans to the arid savannahs of the popular hunter/warrior narrative, for they emerged in a landscape of grassy woodland and continued to have access to forests and rivers and lakes even in steeper and drier areas. With hands freed by upright posture, they used tools to extract matter from plants and scavenged animals, including bone marrow, although some of the marks may have been made by other animals.

The ability to exploit diverse food resources broadened *Homo*'s spatial range and allowed for hundreds of thousands of years of coexistence with partially tree-climbing species of thick-jawed *Paranthropus*, which specialized in tough plants and grasses, and more gracile *Australopithecus*, which ate more fruit and tubers, and some animal foods. Skeletal remains found in the lush area known as the Cradle of Humankind in South Africa starting 2 mya include *H. erectus*, *P. robustus*, and *A. africanus*. Footprints indicate that *H. erectus* lived in an area in northern Kenya alongside *P. boisei* and two other species 1.5 mya (Hatala et al. 2024). *Homo* had a comparatively flatter face, smaller U-shaped mandible, and smaller teeth, thanks to an omnivorous diet.

By convention, eras in human evolution are classified according to styles of stone tools. While a consequence of the durability of stone relative to other materials, this practice exalts a manifestation of cognitive advancement that is culturally, if erroneously, coded as both male and linked to hunting (see Chapter 6). The Paleolithic (Greek for "Old Stone") begins 2.8 mya, the age of the hammerstones and chipped or flaked "Oldowan" (or "Mode One") sharp stones found in Gona, Ethiopia. These tools are attributed to *Homo*, although the same types of tools for cracking nuts or bones and scraping plant and animal tissues have been found at sites occupied by *Paranthropus* in Kenya 3 to 2.6 mya (Plummer et al. 2023) and *A. afarensis* in Ethiopia's Afar Depression 3.4 mya (McPherron et al. 2010). Oldowan tools were succeeded by Acheulean bi-faced flaked hand axes around 1.7 mya, the start of the Middle Paleolithic. These tools were succeeded by the more complexly crafted Mousterian stone tips flaked from a prepared core, used by humans and Neandertals from 200,000 to 50,000 years ago. The Upper Paleolithic, beginning around 50,000 years ago, coincides with the appearance of an intermediate set of artifacts (Initial Upper Paleolithic) for the first few thousand years, followed by the Aurignacian yet-more complex stone tools as well as notation systems, decorative objects, and musical instruments (see Chapter 6). A transitional period starting around 12,000 years ago – before the widespread adoption of agriculture – has the name Mesolithic ("Middle Stone"). This is followed by the "New Stone" Neolithic, and, after that, the Copper, Bronze, and Iron Ages. The dating of these more recent periods varies by region. In addition, stone and bone tools and other implements define specific populations and their time periods, such as Gravettian in Europe and Clovis in the Americas.

In geological terms, the Paleolithic corresponds with the Pleistocene (Greek for "Most New/Recent)" or Ice Age, which spanned from 2.6 mya to 11,700 before the present (BP) and was marked by a series of glaciations. The Holocene (Greek for "Entirely New/Recent") describes a warm era from 11,700 BP to the last couple of centuries, with the start of the Anthropocene. The impact on humans of climatic events corresponding to the Early, Middle, and Late Holocene is discussed in Chapter 7.

Anatomically modern *H. sapiens* emerged more than 300,000 years ago and quickly spread throughout Africa, 100,000 years earlier than previously known from skeletal remains found at Omo Kibish and Herto,

Ethiopia. The current dating is based on human fossils from 315,000 BP ± 34,000 BP at the Jebel Irhoud site in Morocco together with other locations in Africa (Hublin et al. 2017). Fossils dated to 773,000 BP in Casablanca, Morocco, suggest a common ancestor of humans, Neandertals, and Denisovans resembling *H. antecessor* and confirm its presence in the same area where modern humans emerged (Hublin et al. 2026). In Olorgesailie Basin, Kenya, *H. sapiens* left objects 320,000 years ago that included blades and points of chert and obsidian obtained from locations up to 90 kilometers away, as well as decorative pigments from red ochre and black rock (Potts et al. 2018). Groups of *H. sapiens* began to migrate out of Africa at least 180,000 years ago, based on fossils found in Misliya Cave (Mount Carmel), Israel, and reached several Ionian and Aegean islands starting 130,000 years ago and eastern Eurasia 113,000 to 80,000 years ago (see Hershkovitz et al. 2018).

The period around 300,000 years ago was a time of intense wet-dry cycles and shifting food resources that likely spurred *H. sapiens* to make a technological leap to more complex stone tools, both in terms of form and function. Before then, for some 1.4 million years, human ancestors had made simpler tools through reinvention, whereas the new implements required teaching. Cultural transmission suggests improved language ability, which was made possible by a long series of earlier developments (Hage 2024). Humans share speech-related genes and neuroanatomical and neurophysiological features with birds and many mammals. Upright posture detached control of respiration from locomotion. Beginning 500,000 years ago, there were changes to the jawbones, larynx (descended such that air is pushed principally out the mouth), foramen through which the hypoglossal nerve passes, brain, and eyes (small irises surrounded by whites that facilitate nonverbal communication). These changes permit a wide range of sounds compared to primate vocalizations, allowing humans to discuss imagined scenarios, the future, and the past. Nonhuman primates seem not to communicate this way despite being able to put together words they have learned through sign language or image-based communication (see Jarvis 2019), although at least one trained bonobo is able to respond to verbal prompts by representing objects that do not exist, suggesting that imagination has ancient roots (Bastos and Krupenye 2026).

Physical features that distinguish archaic *H. sapiens* from other species of *Homo* include a narrow pelvis, globular cranium, flat face inclined inward at the bottom, smaller teeth and jaw, and brow ridge with central and lateral parts. The skull found in Morocco, as well as a similarly dated skull from China (the Dali fossil), suggest that a less protruding face and not yet fully modern cranium could have been shared with the common ancestor of humans, Neandertals, and Denisovans (see Welker et al. 2020). Humans would outlast these other species thanks in large part to behavioral differences such as larger group size, which facilitates social cooperation, shared parenting, higher birth rates, and group survival. Many genes differed between humans and these archaic species, including genes associated with cognition (see Ju et al. 2025), and modern humans were more genetically diverse (Prüfer et al. 2017).[1]

All populations today descend from the fully modern humans which arose 80,000 to 60,000 years ago in East Africa, where the oldest skeletal remains have been found. Genomic analysis supports this location as the birthplace of modern humans. For instance, a mitochondrial DNA haplogroup (L3) in East Africa has been dated to 80,000 BP and is the likely basis for two mtDNA haplogroups (M and N) present in South Asia 65,000 to 60,000 years ago and which has been found in all genomic samples from people outside Africa.[2] Alternatively, a southern origin is possible, based on the fact that foragers in southern Africa have the highest amount of genetic diversity on the planet (López et al. 2015).

Fully modern humans left Africa from the northeast and spread from the Middle East westward to Europe and eastward across Asia and on to Oceania beginning 65,000 to 54,000 years ago (see Hublin et al. 2020), although they may also have taken southern route through the Arabian Peninsula (see López et al. 2015). They overlapped in time and space with Neandertals and Denisovans, and they overlapped in time with isolated species such as *H. floresiensis*, on the Indonesian island of Flores from 100,000 to 60,000 years ago, and *H. luzonensis*, which lived until 80,000 to 50,000 years ago on the island of Luzon in the Philippines (although this species may be a version of *H. erectus*). By 40,000 BP, Neandertals were gone, and by 20,000 BP only humans were left.

*H. sapiens* reached Australia 60,000 BP from Indonesia, where there were fishing villages 45,000 years ago that were later submerged by rising sea levels; the Americas from Siberia 30,000 to 16,000 BP across the Beringia land bridge and possibly the Pacific; and Madagascar from Africa or eastern locations across the Indian Ocean 10,500 BP (see Ardelean et al. 2020). They had the seafaring skill to push through the formidable current between Taiwan and Japan 35,000 years ago. In the Americas, they used coastal routes and inland rivers to spread rapidly across North and South. Humans also ventured into high-altitude zones in Africa and Asia. They lived at high altitude in Ethiopia as early as 60,000 to 50,000 BP and arrived in northern China more than 40,000 years ago. Humans reached the high Tibetan plateau 40,000 to 30,000 BP with the benefit of Denisovan gene variants beneficial for living at high altitude (such as *EPAS1*). These and other favorable genes for high altitude are widely dispersed among Himalayan populations today and indicate permanent settlement of the plateau 10,000 years ago (Arciero et al. 2025; Zhang et al. 2022).[3] Groups of humans returned to Africa starting several thousand years ago, leaving a mark on the genomes of contemporary populations, especially in Southern Africa and the Somali Peninsula, where Eurasian ancestry reaches up to 40–50% (see López et al. 2015).

## Interbreeding

Over the course of human evolution, hybridization events between species provided a source of new genetic material, one that has not been available since the disappearance of Neandertals and Denisovans. Within the past 124,000 years, *H sapiens* interbred with an archaic species that had branched off before the split between modern humans and Neandertals and Denisovans from a common ancestor, leaving genetic variants in contemporary West African populations that are not found in Neandertals or Denisovans and which account for 2–19% of their ancestry (Durvasula and Sankararaman 2020). Neandersovans migrated from Africa into Eurasia 700,000 to 500,000 BP, having already separated from the common ancestor with humans (Rogers et al. 2020). They eventually replaced the existing Eurasians and diverged into Neandertals in the west and Denisovans in the east, although the two species interbred as recently as within the last 120,000 years. To illustrate, genomic analysis of a more than 50,000-year-old bone fragment found in a Siberian cave indicates that the female's maternally inherited mtDNA was all Neandertal, whereas her nuclear DNA points to a Denisovan father with partial Neandertal ancestry (Slon et al. 2018).

Both Neandertals and Denisovans interbred with *H. sapiens* soon after the latter arrived in Eurasia around 50,000 years ago. Identical gene fragments from related individuals in Eastern Europe reveal both interbreeding and the dispersal of relatives over large distances, for the shortening of these fragments with each generation permits estimation of the degree of relatedness. Based on 45,000-year-old fragments of a skull and spine found in the Czech cave, Zlatý kůň, researchers have determined that this female was related within six generations to a mother and an under-five-year-old daughter whose bone fragments were found in Ilsenhöhle, Ranis, Germany, 230 kilometers away. The mother and daughter's bones were among those of six individuals found in the same cave, one of whom was a female second- or third-degree relative of the mother, while the other three were unrelated males. They lived 49,000 to 42,000 years ago. DNA analysis suggests that over 2,000 years or 80 generations (29-year generation length) had passed since these people's ancestors had mated with Neandertals during the admixture episode 49,000 to 45,000 BP that predated later admixture with Denisovans (Sümer et al. 2025).

Genomic studies have found traces of Neandertal DNA in every contemporary population that has been analyzed, including in Africa. More than 94% of the Neandertal DNA in African populations is shared with populations outside Africa (Chen et al. 2020). At the same time, there are Neandertal and Denisovan variants in Africa that are not found in Eurasia. These came not from admixture upon return migration but rather represent ancestral genetic variants that have been retained in African populations (Bergström et al. 2020). Neandertal DNA makes up less than 1% of African genomes, compared to 2–3% of European and Asian genomes (with more among East Asians relative to Europeans). The genomes of current populations of East Asia and Oceania contain 2–4% Denisovan DNA, picked up by modern humans as they migrated eastward from Eurasia. All combined, Neandertal and Denisovan DNA account for 2–7% of genomes outside sub-Saharan Africa. Contrariwise, modern-human genetic sequences make up 2.5–3.7% of Neandertal genomes, obtained through two pulses of interbreeding in Eurasia, one over 100,000 years ago, the other 65,000–45,000 years ago (Li et al. 2024).

The archaic DNA that remains in the human genome appears to have served a purpose, such as enhanced immune functioning. For instance, DNA from precontact and contemporary indigenous Americans indicates that New World variants of the *MUC19* gene affecting mucus production originated with Denisovans, passed to Neandertals, and was acquired by humans before they migrated across the Beringia land bridge (Villanea et al. 2025). Given the importance of mucosal immunity as the first line of defense against respiratory and gastrointestinal infections, the variants may have provided a benefit to the migrants. The fact that much of the archaic DNA in the human genome is in the major histocompatibility complex likewise suggests that their retention was linked to infectious diseases as a selective force in human evolution given that the MHC contains human leukocyte antigen (HLA) genes for molecules which prompt T-cells to initiate an immune response.

Throughout the Pleistocene, humans migrated and met up in huge numbers at sites where they gifted or traded goods, took advantage of seasonal food resources, and created art and monumental structures (see Chapters 6 and 7). Migrations continued into the Holocene, resulting in continuous gene flow and the dispersal of relatives over long distances. Using the identity-by-descent method mentioned above on more than 4,000 ancient genomes, researchers have uncovered genetic links among widely dispersed Eurasian populations, including the Corded Ware population that occupied a vast area encompassing northern, central, and eastern Europe, and the Afanasievo of the central Asian Altai Mountains. Two Afanasievo individuals buried 1,410 kilometers from each other in Southern Russia and Central Mongolia 3,000 BP were fifth-degree relatives (Ringbauer et al. 2024). The Yamnaya people from the Russian steppes reached eastward to the Altai Mountains and westward as far as Slovakia, covering a 5,000-kilometer-wide area (Anthony 2023; Haak et al. 2015; Racimo et al. 2020).

The population history of Rome over the past 12,000 years provides a fine-grained illustration of gene flow in ancient populations. During the earliest phase, the resident foragers were similar genetically to other European foragers. Thereafter, agriculturalists from Anatolia spread throughout Europe (see Chapter 7), bringing a mixture of genes to the area's residents. Later, between 900 and 200 BCE, the growing city attracted people from North Africa and the Middle East. The population's genomes combined markers typical of those regions with ones similar to markers in current Italians. Genetic diversity in the city's million-plus residents increased dramatically during the centuries of empire (27 BCE to 300 CE), before falling again after the empire's eastern capital was established in Constantinople. The decline of the empire and its western capital shrunk the city's population and opened the door to invasions from the north, resulting in the loss of genetic links to African and Middle Eastern populations and a more uniform genetic profile (Antonio et al. 2019).

In sum, continuous migrations and gene flow have resulted in the broad dispersion of most alleles throughout the world. While some variants are more frequent in populations that have remained relatively isolated thanks to geography or marriage customs, there are no variants present in all members of one population but absent everywhere else. After only 3,000–4,000 generations since modern humans migrated out of Africa, plus a history of movement and mixing, humans everywhere have been left with very similar genomes.

## Quantifying human variation and ancestry

As a result of migrations generally occurring over shorter than longer distances, genetic variation is almost all clinal; that is, gene frequencies shift continuously along a gradient aligned with geographical space (Templeton 1998). Consequently, people are more genetically similar to those around them than farther-distant populations with whom they may be grouped racially or culturally (Handley et al. 2007). For instance, Y-chromosome markers indicate that Muslim men in Andhra Pradesh, India, are genetically closer to non-Muslims in the same state than men of their same ethno-religious group elsewhere on the continent (Aarzoo and Afzal 2005).

Genetic diversity declines with distance from eastern and southern Africa, such that the indigenous populations of Americas, the Far North, Australia, and Papua New Guinea are the least genetically diverse. Migrants take only a portion of the available alleles with them, bringing a restricted pool of variants and less heterozygosity (founder effect). Over time, some alleles become more frequent or "fixed" due to chance events in mating, mortality, and genetic recombination that create a cumulative effect (genetic drift), and there is an increase in the frequency of nonrandom co-occurring non-adjacent DNA segments inherited together (linkage disequilibrium).

Despite these processes, humans everywhere remain nearly identical genetically. Fewer than 0.1% (1 in 1,000) of 3 billion nucleotide bases vary between any two people. Two people living on different continents are only slightly more divergent genetically than any two people living in the same region. In contrast, chimpanzees and gorillas are sufficiently variable to be divided into subspecies (common and bonobo, and mountain and lowland, respectively).

Using a variety of methods, researchers have consistently found a very small proportion of genetic variation between populations (5–20%) compared to within them (80–95%), as summarized by Brown and Armelagos (2001) and illustrated by Figure 3.1. That is, the individuals within a population are far more variable amongst themselves than their population is different from other populations. Studies from before the genomics era examined proxies for genes. Richard Lewontin's (1972) analysis of 17 blood-group, red-blood-cell-enzyme, and serum-protein variants demonstrated that 85.4% of the variation was between individuals within populations. The remainder occurred between populations (6.3%) and between subpopulations within them (8.3%). In the 1990s, using techniques such as the polymerase chain reaction, it became possible to compare bits of DNA. Taken together, eight studies published between 1972 and 1998 yielded the following ranges: 80–91% of the variation within populations; 3–22% between populations; and 2–8% between subpopulations. Proportions at the low end of the range for within-population variation were found in analyses of markers on mtDNA and recombining regions of the Y chromosome, none of which affect morphological, cognitive, or behavioral traits.[4]

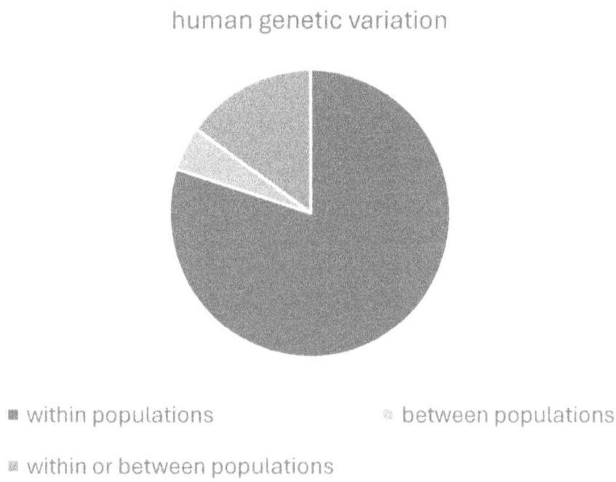


*Figure 3.1* Proportion of genetic variation within versus between populations.

Genomic analysis tells the same story. Rosenberg (2011) analyzed nearly 800 microsatellites dispersed across all the autosomal chromosomes. Microsatellites consist of up to dozens of repeats and are subject to high mutation rates given that they are nonfunctional variants that occur in noncoding DNA. The chosen sequences of base pairs were repeated 2–5 times and had up to 12 alleles each worldwide. Rosenberg's analysis grouped over a thousand genomes from 53 populations into 7 regions: sub-Saharan Africa, Middle East, North Africa, Central and South Asia, Europe, Oceania, and the Americas. The average proportion of differing alleles fell from the first to the last. About half of all alleles were found in all seven regional populations. Almost all were found in at least two (92.5%), with only a small fraction (7.5%) limited to one. Of the latter, more than half were in Africa (56.9%). All the regions contained most of the alleles, with Africa representing the high end of the scale and the Americas the low. The allelic variation across populations came mainly from alleles present in all populations, and the bulk of it, 93.9%, occurred within regional populations, with 3.8% between regions and 2.4% between sub-regions. In a five-region analysis in which Europe, the Middle East, and Central and South Asia formed a single region, the proportions were 93% within regions, 4.5% between regions, and 2.5% between sub-regions.

The biological race concept also fails the test by which biologists classify a population of organisms as a subspecies. Subspecies form where there is altered feeding and mating behavior, and limited gene flow with the parent population. One of Sewall Wright's F statistics, FST (fixation index), compares the variance in the presence or absence of a specific allele in a randomly mating subpopulation relative to the total population. Values range from 0 to 1, with 1 being the strongest indicator of separateness and 0.25 the minimum for consideration in the case of higher-order species. Analysis of several thousand single nucleotide polymorphisms (SNPs) in the 1000 Genomes Project yields a global average FST value between 0.052 and 0.083. Only 10% of loci exceed 0.18; 6.5% surpass 0.250 (Graves 2015; Lala and Feldman 2024). Of the 5% of SNPs with values greater than 0.50, most are recent in origin and found in only one continental population (see 1000 Genomes Project Consortium 2015). Most are in noncoding regions and neutral in effect, with no known functional relevance or relationship to behavior or appearance. In other words, their distribution is unrelated to the traits associated with race.

Other studies concur with these low FST values, which mainly fit within the 0.05–0.15 range. Values are especially low in all of Eurasia (Europe, Middle East, Central/South Asia, plus areas with immigrants from these areas), where pairwise FST values fall below 0.02, with some below 0.003.

Values increase but remain well below the threshold in pairwise comparisons involving populations in southern Africa, the Americas, East Asia, and Oceania (see Kitada et al. 2021; Scott et al. 2016). For instance, a recent analysis yielded FST values from 0.0016 to 0.0030 between European genomic samples (Iberians, Tuscans, and people in Utah with northwest European ancestry), and a value of 0.137 between them and Yoruba samples (Kassem 2025). By all measures, then, there is very little genetic subdivision within the global population.

## Gene-culture coevolution

The human species is unusual not only for its genetic uniformity but also, and related to it, the ability to occupy vastly different environments with diverse climatic conditions. Mutations spread slowly in a long-lived species, and there is none of the non-sexual gene swapping of bacteria. Instead of relying on genetic change, a species with a large, flexible brain comes up with new ideas which can be transmitted immediately through learning and imitation and passed on across generations through cultural processes. As shown below, behavioral solutions such as dietary strategies can themselves act as selective forces upon biological characteristics that become more widespread over time.

In addition to the neutral variants discussed above, there are many functional variants that differ between populations, although not in the way predicted by socially defined racial categories. These have arisen over the course of human evolution, including just the last several thousand years, as a result of selection on traits that confer a benefit in a particular environment (see Hancock et al. 2011). For instance, alleles for straight hair that retains heat are common in cold regions; those for curly hair that disperses heat are prevalent in hot climates. Thick hair and densely packed sweat glands are beneficial in humid zones. Short stature is advantageous in hot climates thanks to a greater surface-to-volume ratio. Alleles prevalent among indigenous Siberians allow heat to be generated directly from fat instead of muscle movement and limit heat loss through reduced shivering and blood vessel constriction.

Reduced skin pigmentation results from many different genes – including *BNC2* inherited from Neandertals – that are common in latitudes above 38 degrees where sunlight is scarce during the winter months. Paler skin more effectively absorbs ultraviolet radiation. Sunlight is needed to synthesize vitamin D from a precursor molecule in the skin and therefore to regulate calcium levels once the liver and kidneys convert calcium into its active form, which supports skeletal, immune, nervous, and cardiovascular system development and functioning. Low-pigmentation genes are typical of but not limited to northern populations. For instance, a common allele among Europeans, *SLC24A5*, has been identified in one-third to one-half of light-skinned KhoeSan peoples of southern Africa, likely introduced through return migration 2,000 years ago and retained because that part of the continent receives less sunlight (Lin et al. 2018). Among multiple variants associated with lighter hair and eye color, a variant of *MFSD12* in East Asians and Native Americans has been found in the genomes of people with high indigenous ancestry in Mexico, Colombia, Chile, Peru, and Brazil, challenging the assumption that lighter skin reflects European ancestry (Adhikari et al. 2019).

The domestication of sheep, goats, cattle, and camels beginning 9,000 to 6,000 years ago has brought the spread of variants of the *LCT* gene (such as *-13910C/T* in northern Europeans or *-13915T/G* in Africans) that cause the cells lining the intestine to continue producing the enzyme, lactase, beyond early childhood (see Holden and Mace 2009). Being able to digest milk sugar is beneficial because it facilitates the absorption of dietary calcium; it is especially helpful to northern populations, who, besides low sunlight, live in an environment where plants grow on calcium-poor soil. Lactase-persistence alleles are prevalent in northern India, the UK, northern Europe, parts of northeast Eurasia including northern China and Russia, and among descendants of these populations in the Americas, Australia, and New Zealand. Lactase persistence is also common among herders in the Middle East and Africa, including Bedouin Arabs, the Tuareg of the Sahara, and the Fulani of the West African Sahel. While lactase persistence has long been considered a case of positive selection, it now appears that the alleles spread in bursts through negative selection against malabsorbers during times of food scarcity or increased infectious disease (Evershed et al. 2022).[5] Dairying has meanwhile altered cattle genetics, for breeds in north-central Europe and elsewhere carry variants that result in earlier weaning and permit sooner milking, along with milk proteins that suit human tastes (see Beja-Pereira et al. 2003).

Humans have been infected by malaria parasites for thousands of years. The transition to agriculture brought dense settlements, cleared land, and pooled water that allowed mosquitoes to thrive and feed on humans instead of other animals. The geographical distribution of diverse variants that reduce the severity and lethality of infection by altering red blood cells (where the plasmodium spends part of its life cycle) overlaps with the regions that have a long history of malaria, especially the most severe strain, *P. falciparum*: sub-Saharan Africa, West Africa, the circum-Mediterranean, the Indian sub-continent, and southern and eastern Asia. Endemic malaria has also favored the spread of a mutation that is protective against *P. vivax* and *P. knowlesi* and nearly universal in West and Central Africa, where these strains are common. The variant (*FY-null* or *FYa-b-*) is virtually absent from Europe, other than areas once occupied by

Moors in southern Spain. It causes red blood cells to lack both types of Duffy cell-surface antigens that bind immune-system-regulating cytokines and are also used by two species of plasmodium, *P. vivax* and *P. knowlesi*. The variant consequently protects against infection with these strains, although it also increases susceptibility to *P. falciparum* by inhibiting immune responses.

Several blood-related diseases protect against malaria but cause serious and potentially fatal disease in homozygotes, including sickle cell disease, one of the monogenic conditions often described in racial terms by biology textbooks. People with two abnormal alleles of the *HBB* gene that codes for β-globin protein produce defective hemoglobin that carries less oxygen and causes red blood cells to mis-form. The combination of sickled shape and reduced survival time results in the harmful and painful accumulation of red blood cells in blood vessels. Heterozygotes produce a mixture of sickled and normal hemoglobin, which attenuates these symptoms while also reducing the availability of red blood cells hospitable to malaria parasites. Other malaria-related balanced polymorphisms include α- and β-thalassemias and glucose-6-phosphate dehydrogenase deficiency, which damage the parasite through increased oxidative stress inside red blood cells (see Etkin 2003). The latter is X-linked and known as favism because of the severe hemolytic crisis that can occur in males upon ingestion of fava beans or other foods high in oxidants. Female carriers may be affected (although less severely) if X inactivation (see Chapters 4 and 5) of some cells allows the defective allele to be expressed. While the extra oxidants cause the erythrocytes of affected individuals to burst, they are otherwise protective against malaria. Oxidant-rich foods such as favas are kept away from boys in malarial regions such as Sardegna, whereas culinary traditions for the rest of the population prescribe serving favas in their most potent raw form during peak transmission season.

Further examples of gene-culture coevolution include multiple copies of the gene for the digestive enzyme, salivary amylase, in South Asian populations whose diet is rich in starchy roots and tubers, and a variety of genetic adaptations to infectious diseases such as deleterious alleles that cause cystic fibrosis but also protect against lung and gastrointestinal diseases (see Hancock et al. 2011). Alleles found among arctic populations protect the cardiovascular system against the effects of a diet heavy in fatty fish and marine mammals (Fumagalli et al. 2015). Tibetans carry alleles that protect against the damaging effects of extremely low humidity and oxygen levels at high altitude (as do their yaks) (Beall 2014).

If races were to be defined by significant, meaningful genetic differences, then Swedes, Bedouin Arabs, northern Italians, and African pastoralists would form a lactase-persistence race, whereas indigenous Americans, southern Italians, and most Chinese people would be a race of lactose malabsorbers. Malaria-protective anemias would make a race of southern Europeans, South Indians, and equatorial African populations and their descendants in the Americas. Black South Africans would align with northern Europeans. Clearly, these groupings do not correspond with traditional racial categories. They do illustrate the non-concordance of traits, which severs the connection in race essentialism between surface features and complex psychological and intellectual traits. Even chromatic characteristics (hair, eyes, skin) are too variable and smoothly graded to allow racial classification, despite their unusual pattern of often varying together. The Humanae Project by the Brazilian photographer, Angélica Dass, provides a powerful illustration in the form of thousands of portraits set against a background matching the Pantone color of each person's nose.

## Genotype-phenotype

The above has shown that no system of racial categorization can accommodate human genetic variation. Gene frequencies fluctuate over time and vary continuously across space. Most of the variation between populations occurs in noncoding regions and is neutral in effect, while the distribution of meaningful alleles does not follow racial lines. The following pages continue the discussion by examining pathways between genotypes and phenotypes to challenge the determinism at the heart of genetic essentialism.

That genes do not determine traits in a permanent and fixed manner is evident from color modifications in a wide range of organisms. The bright color of toxic amphibians depends on the presence of a predator; diverse birds and mammals change color with variations in sunlight between summer and winter; the red primrose (*Primula vulgaris rubra*) flower blooms red at 20 degrees Celsius, white at 30; and flamingoes sport pink plumes if their diet contains carotenoid-rich shrimp (McCartney 2023; Schmid et al. 2022). Sometimes, plasticity combines with genetic changes that become widespread in a population. For instance, side-blotched lizards in the Mojave Desert that live in sandy areas are brown and tan in color, while those on the Pisgah lava flow are black. In the lab, switching them results in color shifts within a few weeks. However, the color of the lighter lizards placed on a dark surface does not fully reach the dark color of the regular lava dwellers. The latter carry two mutated genes (*PREP* and *PRKAR1A*) that code for melanin production, illustrating the genetic assimilation of a plasticity-generated darkening that improved the survival of ancestors which first stepped onto the cooling lava flow 22,500 years ago (Corl et al. 2018).

Even the classic examples of Mendelian inheritance do not follow a simple dominant-recessive pattern. The pea-plant traits Mendel described as dichotomous appear along a range of values, especially outside controlled laboratory conditions (Dougherty 2009). In humans, earlobe attachment may be more or less complete and may differ between an individual's two ears, while eye color involves incomplete dominance and more than a dozen genes. Similarly, the monogenic diseases typically used as racialized examples in textbooks yield a range of symptoms, such as sickle cell, noted above, and cystic fibrosis, which stems from several different recessive alleles of the *CFTR* gene. The associated protein is involved in the transmembrane passage of ions, without which mucus builds up in the lungs and elsewhere. Homozygotes with two copies of the same mutation are rare. Many more patients carry two different deleterious alleles, which creates a range of symptoms from mild to severe. Dozens of other monogenic diseases likewise do not manifest themselves uniformly across patients, including Gaucher disease, Charcot-Marie-Tooth disease, and certain forms of congenital adrenal hyperplasia (see Dipple and McCabe 2000). As noted in the last chapter and illustrated in Figure 3.2, people born with two recessive alleles of the *PAH* gene that affect the phenylalanine-to-tyrosine conversion enzyme, phenylalanine hydroxylase, need not develop phenylketonuria (PKU) at all: dietary restriction of phenylalanine plus tyrosine supplementation prevents phenylalanine from building up and damaging the brain.

Moreover, rare disease-causing alleles are not restricted to the populations identified by racialized examples in textbooks. For instance, mutations of the tumor-suppressor genes *BRCA1* and *BRCA2* that increase the risk of breast (and ovarian) cancer are distributed throughout the world, not only among Ashkenazi Jewish women. These variants have a high impact among the affected 2–3 women per 1,000, yet they do not uniformly result in breast cancer.[6] Contrariwise, absence of these genes does not signify absence of disease risk, for 90–95% of cases occur in people with neither mutation (see Mehrgou and Akouchekian 2016).

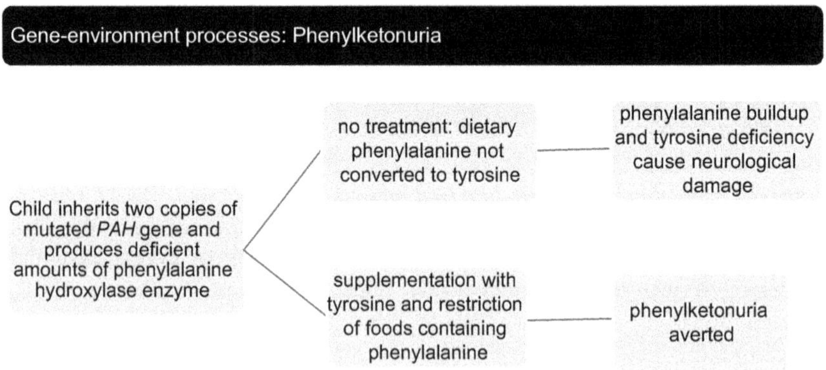


*Figure 3.2* Variability in the expression of a monogenic disorder: Phenylketonuria.

Children's growth and adult stature depend on thousands of genes, along with environmental factors including diet, physical activity, infectious disease, and stress. For this reason, polygenic scores do accurately predict height (within a few centimeters). Children's life circumstances shape how much they grow even in the presence of known height-associated variants. For instance, despite sharing alleles associated with shorter stature – such as variants of *GHR* (growth hormone receptor) and *IGF1* (insulin-like growth factor) – children of Central African foraging populations who are raised in nearby villages grow taller than those who live in the forest (see Becker et al. 2013). Throughout the world, national averages have increased dramatically over the past century or two wherever living standards have improved, just as moving to a wealthier country brings increased height among the children of immigrants – such as a nearly 12-centimeter difference between Mayan children of immigrants to California and Florida and their peers in Guatemala (Bogin et al. 2002). The same changes in nutrition and infectious disease exposure have brought a decline in the "natural" or "normal" human body temperature of around 0.5 degrees Celsius in the US and other wealthy nations, further illustrating gene-environment processes (see Protsiv et al. 2020).

While multiple genes may contribute to a single trait, single genes can influence multiple traits (pleiotropy). For instance, many variants correlated with elevated risk of schizophrenia also increase the risk of inflammatory bowel disease; variants associated with later menarche in women correlate with multiple traits including delayed voice drop and lower risk of pattern baldness in men (Pickrell et al. 2016). Moreover, genes at separate loci may suppress each other's expression or combine to yield a different trait, adding another element to the way alleles may manifest differently across individuals, including diseases caused by a dominant gene such as Marfan Syndrome (see Wray et al. 2018). These epistatic interactions upset the 9:3:3:1 ratio of a Mendelian cross between parents heterozygous for two traits (see Chapter 2). To illustrate, genes for pigment production in fruits and flowers can be masked by other genes. In summer squash, a gene for white fruit silences a separate gene for colored fruit, resulting in a 12:3:1 ratio: 12 combinations resulting in white fruit (any pair in which there is at least 1 dominant *W* allele), 3 combinations for yellow fruit (*wwYY* or *wwYy*), and 1 combination for green fruit (*wwyy*). Albinism is similar although it involves diverse forms and is due to recessive alleles of several genes that interfere with the production of pigment normally regulated by multiple other genes.

The importance of noncoding DNA and other factors to phenotypic expression has become increasingly clear since publication of the first fully sequenced human genome by the Human Genome Project in 2003. At the time, the molecular model predicted an equal number of genes to proteins. Instead, the number of genes turned out to be many times smaller than proteins: around 20,000 versus 100,000. In addition there are thousands of nonstandard coding segments or microgenes, which lack the usual start and stop sequences yet are transcribed and translated into functional proteins with as few as 12 instead of the usual 100+ amino acids. All told, genes account for less than 5% of the 3 billion base pairs in the human genome, leaving at least 95% of the genome as noncoding. That humans share 98% of their genes with chimpanzees and gorillas, and nearly as many with other species including mice, points to the role of the noncoding regions in regulating genetic expression and accounting for species differences. Together with epigenetic patterns discussed below, noncoding regions explain how cells carrying the same DNA can exhibit diverse, context- and developmental-stage-dependent forms and functions.

The image of DNA as a blueprint is highly misleading given that transcription and translation depend on the presence of cellular components and the energy generated by proteins, which vary with location in the body and a host of environmental factors. Moreover, the 37 genes in human mitochondria affect the expression of nuclear DNA, while hundreds of proteins from nuclear DNA influence the expression of mtDNA. Proteins likewise affect each other. They pass information independently of DNA, as happens when prions cause normal proteins to misfold in the brains of people and animals with transmissible spongiform encephalopathies. In addition, extrachromosomal DNA – circular loops present in many cancer cells (and, in much smaller form, some normal cells) – carry genes that favor tumor growth, impede anti-cancer immune functions, and increase resistance to medicines.

Different genomes in the same individual may arise through alterations to DNA during cell replication, bringing genetic differences between types and groups of cells, such as skin, blood, and brain cells. Even the number of chromosomes may vary, with some cells carrying too few or too many, as occurs regularly in heart, liver, and bone marrow cells. These phenomena suggest that genetic redundancy and variability serve a purpose, such as resistance to stress, tissue regeneration, and structural support. In the lab, cells that have already differentiated may be coaxed into reverting to pluripotent cells, and one cell type can be converted to another.

The activities carried out by RNA inside and outside the cell add another layer of complexity. Besides DNA transcription and translation, RNA molecules trigger immune responses, orchestrate cell signaling, and prevent errors such as the translation of faulty mRNA into proteins or the transposition of DNA fragments from one locus to another. RNAs generated from noncoding regions regulate genetic expression. Hundreds of proteins can result from the same gene due to the way mRNA splits DNA into pieces and rearranges them during transcription, as occurs in many cell types including B lymphocytes, brain, and skin.

Genetic expression is shaped also by epigenetic modifications that do not affect DNA itself. Epigenetic patterns serve many normal regulatory purposes, including suppressing the expression of defective DNA sequences. They are stable across cell divisions but may be modified and reversed in response to novel exposures. Epigenetic patterns differ by type of cell and tissue. They work through various mechanisms, such as the attachment or removal of a methyl (hydrocarbon) group to or from the outer reaches of a nucleotide base pair (most often cytosine-guanine), and the binding of tiny RNAs to DNA that alters its expression during transcription or translation. Another involves the histone proteins that form the core around which the DNA strand is coiled, each nearly two-turn twist around eight histones constituting a nucleosome. During transcription, while a nucleosome is partially unwound, methyl or acetyl enzymes may bind to or detach from a histone in ways that either activate or suppress gene expression.

Epigenetic patterns explain how genetically identical social insects such as bees and ants become anatomically and behaviorally divergent queens and workers (and, in the case of carpenter ants, subdivide further into castes specialized in either foraging or defense). In humans, they play a role in the emergence of different phenotypic traits between monozygotic twins. Environmental and experiential conditions that cause epigenetic changes in humans and other organisms include nutrients (injurious types and/or quantities), toxins, trauma, psychosocial stress, altered hormone levels, parental neglect, and social isolation (and their opposites) (see Graves 2015). In both laboratory animals and humans, inadequate maternal care induces epigenetic modifications that result in increased reactivity to stress in adulthood. Altered epigenetic patterns have been found in children of mothers who experienced trauma during pregnancy, such as poverty, war, severe overnutrition or undernutrition, toxic exposures, or other stressful events. Fathers also appear to pass on epigenetic modifications from harmful exposures, although intergenerational transmission of changes in the germ cells remains under investigation and is less likely compared to plants and laboratory animals given that fertilization brings the erasure of many epigenetic marks on the paternal and maternal genomes. Links from stressors to specific health impacts by way of epigenetic modifications include toxic environments to cancer; psychological trauma to neurological changes and reduced resilience to stress; and poor nutritional quality to higher likelihood of metabolic and cardiovascular diseases, with a reduction in life span. In contrast, positive experiences and environmental circumstances alter epigenetic marks and translate into more favorable physical and mental health.

Environmental influences on phenotypic expression also include processes involving genes, mRNA, and proteins that affect development, beginning with the egg's orientation in the ovary and the contents of its cytoplasm.

Mothers transmit their microbiota to babies during childbirth, with wide-ranging impacts on physical and psychological development and functioning. If mice are any indication, during pregnancy maternal microbes pass by-products of digestion to the fetus placentally, affecting its lifelong susceptibility to obesity and metabolic disease (see Kimura et al. 2020). Babies also inherit antibodies in mother's milk that have a long-term impact on the development of their immune system. They inherit ecological and socioeconomic environments, cultural beliefs, and educational and occupational opportunities.

From birth on, commensal microbes that inhabit the gut (bacteria, viruses that infect bacteria, archaea, mycoses) influence genetic expression. Far more numerous than the body's own cells, these microorganisms make up a mass that would rival a large organ in size. They break down medicines and food, and their digestion of fiber generates short-chain fatty acids such as propionate, acetate, and butyrate that have wide-ranging signaling functions beyond the digestive tract. Short-chain fatty acids stimulate the production of regulatory T cells which suppress inflammation. They facilitate cognitive development in infants, protect the intestinal barrier, modulate metabolic functions, promote psychological wellbeing, and regulate social behavior. The *Blautia* genus of bacteria illustrates the vast range of impacts a single type of microorganism may have on the body. These bacteria not only reduce the risk of metabolic disorders and obesity in humans and experimental animals but also protect against the inflammation-mediated effects of hypoxia on the brain and other organs in altitude sickness (Su et al. 2024).[7]

The complexities outlined above make it clear that phenotypic traits do not arise automatically from single genes that code for one protein each and affirm that current health disparities are biologically linked to past conditions, as opposed to separate versions of human biology.

## Genomics, determinism, essentialism

The complexities of phenotypic expression described above complicate the interpretation of statistical associations identified by GWASs and explain why most such associations are weak. While GWASs spotlight polygenicity, have important clinical implications where the mode of action is known or can be determined, and have already brought advances in knowledge of cancer genetics, disease processes, and responses to pharmacological and immunological treatments (see Claussnitzer et al. 2020), they also project a deterministic view of causation and, where structured by racial categories, endorse the biological race concept.

In general, GWASs search for statistical relationships between a trait of interest and some of the over 10 million SNPs in the human genome.

Identifying SNPs narrows the focus to nearby genes through a form of guilt-by-association. That is, mutations which have spread in a population are not likely to have become split from the SNPs close to them through crossover events during meiosis. The associations are mathematical and therefore do not demonstrate causation; moreover, they concern a particular set of samples beyond which they may not be generalizable. This is as much the case for GWASs as phenome-wide association studies that consider variables such as diet, physical activity, microbiome composition, environmental quality, socioeconomic status, and psychosocial stress.

Even genes with a strong association to a trait in one population may be weakly or not at all associated with the same trait in another population. For instance, an analysis of more than 3,000 genomes of US patients with celiac disease in the All of Us (see below) database found that the strongest associations to around 1,600 SNPs occurred in genomes with European ancestry. These SNPs are mainly in the major histocompatibility complex and concern the production of HLAs, mentioned above. One haplotype (HLA-DQ2.5) occurred at lower frequency in non-European-ancestry genomes, while a haplotype block encompassing HLA-DQ2.5 and other elements (including HLA-A1) was absent from them (Karnati et al. 2024).

Genes with a high impact tend already to be known, as in the case of diseases such as sickle cell, Huntington's, and hundreds of rare disorders, thanks to their early onset, severe impacts, and familial transmission, although thousands of additional candidate alleles already identified by GWASs could shed light on the variability in their severity. More commonly, genes associated with a trait are many and only weakly associated with it. To illustrate, dozens of alleles have been linked to stuttering, and some of them also with depression, autism, and disorders of rhythm and movement (Polikowsky et al. 2025).[8] The associations suggest avenues for research that could lead to treatments and are helpful for combatting blame and stigma surrounding stuttering. At the same time, they implicate multiple genes, while the shift in male-to-female ratios of affected individuals from near equality at onset to 4:1 among adolescents and adults indicates that complex gene-environment processes are involved. Furthermore, the alleles whose mode of action is known do not explain how un-stuttered verbal communication is produced.

Common pathologies such as cancer and cardiovascular disease involve thousands of variants, each with very weak effects (see Novembre et al. 2022). Polygenic risk scores represent the sum of these variants, weighted by effect sizes. Most of the disease-associated variants are in noncoding regions of DNA, suggesting that they influence how DNA is expressed. The same causal variant may have divergent effect sizes in different populations due to epistatic interactions, age and sex effects, and gene-environment processes, while noncausal variants may be in linkage disequilibrium with causal variants that remain to be identified and which may not be the same across populations. Disease-associated variants may be compensated by buffering mechanisms that reduce the impact of defects and deficiencies in the production of proteins. The fact that healthy individuals carry dozens to a hundred or more loss-of-function mutations indicates that redundancies and back-up systems can prevail over deleterious genes.

In addition to differing across cell, tissue, and organ type, genetic expression varies across stages of disease and phases of life from embryonic development to older age, as mutations accumulate. As a result, people with the same polygenic risk score may exhibit a range of phenotypes. Conversely, people with divergent genetic profiles tend to have a similar likelihood of developing common diseases. Roberts et al. (2012) analyzed data on two dozen diseases (including non-insulin-dependent diabetes, dementia, cardiovascular diseases, inflammatory bowel disease, and nine cancers) from 54,000 identical twins across five countries. The results suggest that, for most of the diseases, not having genetic risk factors has only a modest impact on the chance of developing disease. The risk among those with a negative test remains at least 50–80% of the general population's risk. At the same time, whole genome sequencing turns up genes associated with at least one of the diseases in nearly everyone (over 90%). The authors conclude that genomic information is less useful for preventing disease than existing measures. For instance, with respect to heart disease, known risk factors such as lipid levels, blood pressure, smoking, and age are far more predictive.

While the frequency of rare alleles varies between populations, most common variants are widespread throughout the world and occur across racial/ethnic populations at roughly equal rates (Graves 2015; Wray et al. 2018; Visscher et al. 2021). This uniformity highlights how health disparities are rooted in factors such as environmental exposures, the harmful effects of experiencing racism, and inequalities in medical care. In fact, the frequency of homozygosity for harmful alleles is lower among Americans of African than European ancestry, which, if genes were all that mattered, would confer better health, not worse (Lala and Feldman 2024). An analysis of genetic ancestry test results of individuals who identify as African American in Cleveland, Ohio, indicated predominantly European ancestry in 96% of cases (half or more of genetic markers), whereas African ancestry was indicated in 4% of cases. Yet, despite sharing genes with European Americans, African Americans in Cleveland and elsewhere do not share European Americans' disease rates, experience of illness, or outcomes (Graves 2015).[9]

Health disparities, rather than unspecified genetic peculiarities, are the reason genomics researchers divide data by population. An analysis of over 200 peer-reviewed scientific papers published from 2003 to 2017 investigated race either as a social construct, in relation to outcomes, or as a genetic category (Duello et al. 2021). This last group accounted for about 60% of papers. Their rationale was epidemiological rather than genetic, such as a difference in heart failure mortality between middle-aged Black and White people (see below). The studies identified African-ancestry participants by way of self-reported race or skin color, or surrogate-reported race. Of 117 studies that focused on African Americans, only 30 included more than 500 people, and 31 had 100 or fewer participants, yet none of them considered whether the data were generalizable to other African American populations. The corpus of papers reflects a view of African ancestry as fixed and discrete, contrary to the continuous and ever-changing nature of human biological variation. The authors hypothesize that researchers may be under the impression that the purpose of increasing the participation of underrepresented groups in scientific studies is to uncover racial differences in physiological functioning, analogously to the scientifically justified correction to the exclusion of women from medical research.

The heart failure statistic mentioned above justified the US Food and Drug Administration's first approval of an "ethnic drug," BiDil, which combines two already existing medicines, isosorbide dinitrate and hydralazine (Sankar and Kahn 2005). The medicine works in all patients and was not tested for differential efficacy by race. Instead, it was developed based on decades-old data indicating that middle-aged Black patients died of congestive heart failure at twice the rate of White patients. Focusing on middle age exaggerates the racial difference because most mortality from this disease is concentrated in older age groups, especially in the White population. The difference in heart failure rates for all age groups combined was narrower at the time, and by the early 21st century was close to zero (1.08 to 1). Thanks to patent protection, the drug costs seven times more than the generic ingredients, exacerbating the inequalities that generated the difference in mortality rates in the first place.

Focusing on biological race in genomics converts race into a risk factor and shifts attention away from social and environmental factors, as shown by the example of non-insulin-dependent diabetes (Type II), a disease that spreads in every population that switches to a sedentary lifestyle and carbohydrate-rich diet. The discovery of dozens of genetic loci associated with the disease reinforces longstanding assumptions about biology and race (see Duncan et al. 2024). As McDermott (2016) shows in the case of indigenous Australians, these conceptual effects translate into racial discrimination on the part of health professionals, the individualization of the disease, and inattention to the sociopolitical systems that constrain people's health-related behavior and access to healthful food.

Preconceptions about race and disease can also lead health professionals to withhold genetic testing from patients whose appearance is assumed to preclude their carrying the relevant genes. Carter (2021) describes cases in which symptomatic Black patients have been denied testing for genes known to play a causal role in amyotrophic lateral sclerosis, due to the misconception that ALS is a "White disease." Although the disease is currently incurable, genetic testing can rule out diseases with similar symptoms, while not testing symptomatic patients reduces the pool of data on which the scientific community may gain further knowledge that could result in gene therapy or other treatments. This example shows that, for genomic science to serve everyone equally, testing should be based on medical information rather than essentialist assumptions about ancestry.

Equitably distributing the benefits of advances in genomics also requires expanding the pool of genomes from underrepresented groups. Current associations vary in strength across populations, resulting in differential accuracy of polygenic risk scores, especially in the case of relatively isolated groups relative to large continental populations (see Stæger et al. 2014). Most large research and commercial databases are heavily weighted toward middle-aged, economically advantaged individuals of European ancestry. For instance, the UK Biobank contains genomes of 500,000 British people aged 40–69, only 6% (20,000) of whom are of Asian or African descent. Young men are underrepresented, and the pool is skewed toward healthy, relatively well-off people. Efforts to expand the data available for genomic analysis include the China Kadoorie Biobank, with over 500,000 genomes from 10 ethnic groups, and the Human Heredity and Health in Africa consortium, with 120,000 participants from throughout the continent. Three-quarters of the first 245,000 genomes in the US National Institutes of Health (NIH) study, "All of Us," came from groups that are often left out of medical research, including older Americans and people whose socioeconomic status falls at the lower end of the scale. Almost one-half of individuals identified as having non-European ancestry, with slightly more than one-half determined by imputation to be of non-European ancestry. Analysis of these genomes has turned up millions of previously unreported coding (about 4 million) and noncoding (around 272 million) variants. Nearly all of the coding variants (99.98%) are rare, with an allelic frequency under 0.01, whereas half of the common coding variants (0.01%) are rare in the European ancestry group but common among one or more non-European ancestry group (All of Us Research Program Genomics Investigators 2024).

Adding more genomes from underrepresented groups not only leads to a more complete picture of the genetic variability among humans but also challenges rather than confirms the existence of biologically distinct races aligned with contemporary folk categories. For instance, an analysis that included larger numbers of samples from African individuals collapsed the seven continental reference populations generated by the Human Genome Diversity Project and 1000 Genomes Project databases into three. Five of the seven clusters aligned with subpopulations within the African continent, leaving only two non-African population groups. Furthermore, the seven continental populations blended into each other in the BioMe dataset of genomes from New York City residents (Bird and Carlson 2024).

Even though all ancestry is mixed, genomic studies commonly apply a human-made typological scheme to divide samples into population groups and use algorithms to cluster data accordingly. Visual representations arrange populations in a treelike structure which reinforces the idea that current humans descend from distinct ancestral groups. Populations are assumed to be made up of randomly mating individuals, which is never the case. Models also unrealistically assume that genes are inherited independently of one another (no linkage disequilibrium) and that there has been no immigration, selective pressures, or mutations to upset Hardy-Weinberg constancy in gene frequencies (Graves 2015). These unrealistic assumptions make the genetic characterization of populations inherently unreliable.

In contrast to GWASs that search for markers associated with a particular trait of interest, ancestry testing involves comparing portions of an individual genome to reference populations with different frequencies of markers such as SNPs, identical sequences, and haplotypes, especially in maternally inherited mtDNA and the non-recombining portion of the Y chromosome. Percentages and graphics imply that discrete fractions of the genome derive from these populations, while polygenic risk estimates suggest genetic determination of physical traits and disease susceptibilities. The ancestry part is deceptive because the genomes making up reference populations come from contemporary humans, not source populations from the distant past. Moreover, shared genetic markers are not unequivocal evidence of genetic relationships or historical connections, for the same variant may arise in diverse populations through parallel selective pressures or random mutation.

The ancestry-testing industry capitalizes on the essentialist idea that genetic ties to geographical populations, ethno-religious groups, and famous historical figures reveal something eternal and true about the self. However, while it is possible to reliably confirm genetic ties between close relatives thanks to the types of markers involved, the nature of human genetic variation greatly reduces the accuracy of ancestry estimation beyond the continental level. And, beyond the most recent generations, people's DNA comes from a great many relatives, each with a tiny contribution. Five generations back, each of 32 ancestors contributed 3.12% of an individual's genes. Ten generations back, 1,024 ancestors each contributed 0.1% of the genes.[10]

These complexities suggest caution in interpreting not just current but also ancient DNA. While aDNA can help shed light on migrations and relationships between groups, as seen above and in Chapter 7, it is unevenly distributed around the world and over human history due to burial conditions, mortuary customs, and the effects of time. Just as biobanks are unrepresentative of all humankind, so is the aDNA record. And just as contemporary genomic data are collected and interpreted within a sociopolitical context, so are aDNA sequences. Storytelling about the deep past dehumanizes individuals by reducing them to their genomes and sacrifices local detail, complexity, and gaps in knowledge for a grand historical narrative. Current cultural preoccupations with gender and race focus attention on sex typing and skin, hair, and eye color. Visual reconstructions such as those described in the next chapter concretize these themes in an immediate, authoritative way, effacing the uncertainties familiar to genetics and demonstrating the need to diversify not only the DNA record – both ancient and modern – but also the pool of people who narrate it (Källén 2025).

In sum, by continuing to operationalize the biological race concept, ancestry testing and genomic studies together sustain a deterministic view of genetic causation and imply the existence of genetically distinct populations. Chapters 1 and 2 showed that the perception that races differ in disease susceptibility leads to the perception that they differ in cognitive and behavioral traits as well. As shown below, genomics studies on the latter characteristics smooth the path from bio-somatic to bio-behavioral essentialism.

## Genomics, behavior, and intelligence

The collection of personal data along with biological samples for large databanks has opened the door to analyses linking genes to behavior. For instance, the UK Biobank made publicly available in 2017 includes information on more than 2,400 traits, including physical and mental health, lifestyle habits, social life, and education level. Tapping into such sources allows researchers to characterize the "genetic architecture" of behavior, personality, and intelligence by identifying hundreds to many thousands of genes with faint associations to behaviors such as caffeine or nicotine consumption. With respect to the "big five" personality traits, weak correlations have been found between hundreds of markers and neuroticism, as well as far smaller numbers of markers and agreeableness, extroversion, openness, and conscientiousness (Gupta et al. 2024). While these studies may contribute to the search for preventive or therapeutic measures related to addiction or the psychiatric conditions often linked to neuroticism (such as depression and schizophrenia), they sustain a deterministic vision whereby a genetic foundation for cognitive/behavioral traits exists independently of developmental inputs, life experiences, personal identity, or individual agency.

As with gene variants with known effects on disease, those which alter neuronal morphology or signaling do not cause behavior on their own. To illustrate, the monoamine oxidase enzymes encoded by the X-chromosome *MAO-A* and *MAO-B* genes degrade neurotransmitters, including serotonin, dopamine, adrenaline, and noradrenaline, which prevents their buildup in the brain. Whereas men with a variant of these genes have only the one allele, women with a defective allele on one X may still produce appropriate amounts of enzyme if there is a normal allele on the other and it is not silenced by Xinactivation. The link between low-activity alleles, excess serotonin, and aggressive and criminal behavior has led to the idea of a "warrior" or "criminal" gene. However, whether these genes affect behavior depends on the co-presence of adverse life conditions (see Mentis et al. 2021). Abuse during childhood together with low-activity alleles increases the risk of antisocial personality disorder; current stress combined with the alleles is associated with increased crime and aggression, which is especially more likely if there has been early life stress. While high-activity alleles protect against the effects of stressful life experiences, they increase the risk of psychiatric conditions, including anxiety, major depression, and bipolar disorder by way of dampened neurotransmitter levels.

No longer limited to calculating heritability, the behavioral genetics field now draws upon genomics data to produce polygenic scores for intelligence and educational outcomes. A mixed-methods study involving US pre-school, primary, and secondary school teachers suggests that many educators would be on board with efforts to "personalize" educational pathways for children based on their genetic profile (Martschenko 2020). Participants expressed interest in learning about the potential role of genetics in educational policy. They tended to believe that student intelligence and academic performance depend heavily on inherited genes, while the environmental influences they recognized as relevant were principally parents and the family environment.

Besides the ethical and privacy issues it raises, a genetic approach to children's education ignores the complex environmental conditions and structural features that impinge upon children's mental health and academic performance, regardless of their genomic profiles (see Sabatello 2018). In addition, it wildly exaggerates the effect sizes of genetic variants associated with cognitive abilities and academic outcomes. For example, the few relevant alleles identified through GWASs each explain less than 4% of the variation in IQ test scores (Lala and Feldman 2024).

Attributing children's academic ability to genes supports the hereditarian claim that racial differences in educational attainment are the result of natural selection, even though environmental confounding casts doubt on any claim of genetic causation. Instead of random mating, human couples tend to share a variety of behavioral, educational, and social characteristics that create correlations between bundled alleles. GWASs cannot separate covarying environmental and genetic factors between populations (Bird and Carlson 2024; Novembre et al. 2022). The only way for hereditarians to prove their thesis would be to dismantle structural inequalities and create a fully egalitarian society in which intelligence genes could shine through the muddle (Donovan 2022).

Even if intelligence could be accurately defined and measured, and even if society eliminated the cultural beliefs about innate ability that distort performance on tests when people are reminded of stereotypes, basic principles of evolutionary biology exclude the possibility that groups of humans differ in intelligence-related genes. It is not likely that any one of the more than four-fifths of all human genes expressed in the brain could have a large effect, for its familial transmission would have been noticed by now. No evolutionary process could explain population differences in intelligence-related genes, for thousands of genes are unlikely to have varied together through genetic drift, and there is no evidence of unique selective pressures that would favor the spread of different forms or powers of intelligence in some populations but not others (Lala and Feldman 2024). As noted in the last chapter, people in different societies may develop different cognitive abilities, but the cause is culturally defined uses, not separate brains. Factors that contradict essentialist beliefs about innate differences in intelligence are summarized in Figure 3.3.

Indeed, intelligence-associated variants fit the pattern whereby socially defined races vary far too little genetically to be considered different subpopulations. A comparison of polygenic scores related to academic achievement between African American and European American genomes in the 1000 Genomes Project demonstrates that FST values for education-associated variants are too low to indicate genetic differences between groups (Bird 2021). Moreover, the average IQ difference is narrower than claimed by hereditarians.

A GWAS involving 1.1 million samples from the UK Biobank and 23andMe yielded polygenic scores involving 1,271 markers that explained 11% of the variance in educational attainment, and 7–10% of the variance in three cognitive variables: cognitive test performance, highest-level math course completed, and self-reported aptitude in math (Lee et al. 2018). When applied to African American samples, polygenic scores explained only 1.6% of the variance in educational attainment. While the authors expect that more predictive polygenic scores for non-European populations will become available once databases include more genomes from these groups, Lala and Feldman (2024) point out that the reason for the gap between percentages is likely that educational opportunities and attainment fall into a narrower range for African Americans than European Americans.

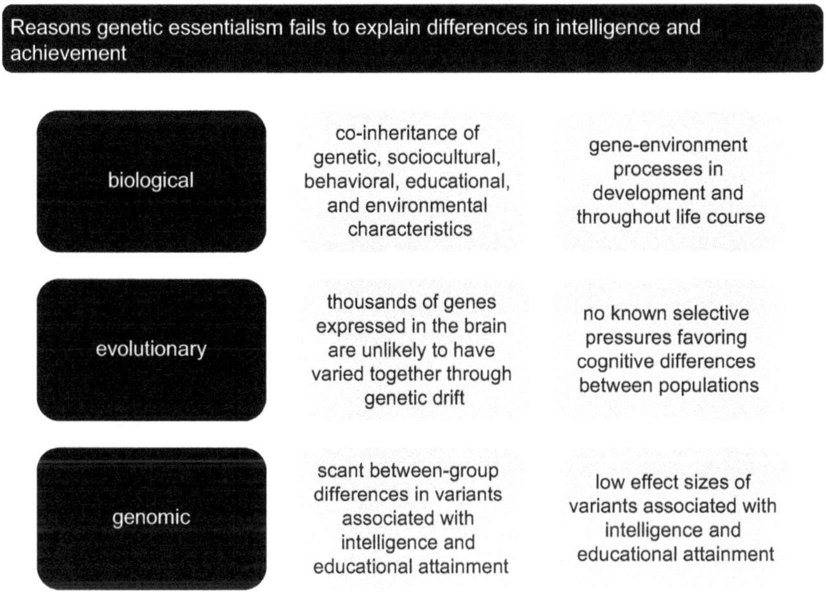


*Figure 3.3* Reasons genetic essentialism fails to explain differences in intelligence and achievement, by category.

The "human biodiversity" field seized upon Lee et al.'s study to support its genetic essentialist vision of intelligence and achievement which, as noted in earlier chapters, undergirds hostile rhetoric and can motivate violence. An image posted anonymously to the online bulletin board, 4chan, showed the education-associated SNPs with the highest frequency difference between genomes of people with European versus African ancestry to insinuate that science backs the notion that African ancestry makes people less intelligent. The meme then circulated across multiple social media sites. It was included in the screed left by the White supremacist who killed 10 people in a Buffalo supermarket in 2022 (see Bird and Carlson 2024).

Besides circulating misappropriated research findings, hereditarians access public biobanks including several maintained by the US NIH to generate their own numbers in support of racist, homophobic, and sexist arguments to spread through social media (see Bird and Carlson 2024). In 2026, an international group claimed a basis for innate White intellectual superiority after bypassing NIH security guardrails intended to protect nearly 12,000 children whose brain scans, biospecimen data including genomic information, and cognitive, social, behavioral, and mental and physical health data were collected for the Adolescent Brain Cognitive Development study. The study, which began in 2015 and follows 9–10-year-olds for 10 years, was designed to include a representative sample of children from multiple sites throughout the US and consequently included self-identified race/ethnicity using US Census Bureau categories (see Karcher and Barch 2021).

To counter the potential for genetic data to be mined for racist ends, the American Society of Human Genetics (2020) has issued a statement against the genetic race concept and the White supremacist idea that there exist "good" genes in some populations but not others. Yet, despite such statements and the anti-racist position of the American Psychological Association, professional journals and book publishers continue to print studies in which authors manipulate concepts and data from evolutionary biology, genetics, and genomics to stamp popular race categories with biologically validity (see Bird et al. 2024).

Beyond the potential for harmful mischaracterization, genomics research raises many ethical and social questions. Easily accessed databases combine genomes from people who have given their samples and consent only for biomedical research or commercial ancestry testing, not purposes that could harm their or other groups. Two years after a massive data breach, in 2025 the 23andMe company's bankruptcy put 15 million people's genomic, personal, health, and address and credit card information up for sale (ultimately to the non-profit TTAM Research Institute). The fact that their information as well as the original biological specimen could end up in the wrong hands may further discourage underrepresented groups from allowing their genomes to be sequenced, which would compound the problem of uneven sample sizes and lend further support to policies and practices that can further marginalize them.

Social consequences of genomics research and personalized medicine include the diversion of resources away from public health and a disproportionate benefit to the wealthy and well-insured (see Tabery 2023). Privacy is under threat given the ease with which DNA can be collected without a person's knowledge and their identity traced regardless of anonymization procedures, especially if the entire genome has been sequenced. Large genomic databases that reliably turn up third-degree or closer relatives can be accessed by people seeking relatives as well as law enforcement investigating crimes (see May 2018).[11] Research on socially consequential characteristics such as sexuality, disability, and smoking and other health-related behaviors could penalize individuals in many ways, including medical and other forms of insurance, and can be expected to occur based on the discrimination against Black Americans after the genetic basis of sickle cell disease was discovered in the 1950s. The 2014 US Genetic Information Nondisclosure Act only protects against employment discrimination and workplace harassment (and only in organizations with more than 15 employees), while the Health

Insurance Portability and Accountability Act does not cover commercial ancestry testing.

Based on GWAS data, prospective parents may demand prenatal screening for a wide range of traits unrelated to health and, some day, prefer assisted fertilization with genetic manipulation for desired characteristics. Arms races for socially valued traits would bring inequalities in access, prejudice against parents (and their children) who opt out, and the loss of an open future for children whose interests and aptitudes have been chosen for them. It would matter little to parents with a deterministic understanding of genetics that the project is destined to fail, given that all traits are 100% nature, 100% nurture.

Genomics studies feed the popular enthusiasm for genetic explanations of complex traits. Those which incorporate a racial typology naturalize the social construction of race and appear to substantiate the misperception that populations have discontinuous physical and mental traits. The following chapters examine how this misperception colors interpretations of sex/gender differences, to the point of introducing erroneous information into science textbooks. The discussion builds upon the above analysis of human migratory history, genetic variation within versus between populations, and gene-environment processes in phenotypic causation, to further demonstrate how genetic essentialism is antithetical to human variation.

## Notes

1 Prüfer et al.'s analysis demonstrating larger group size and greater genetic diversity in humans compared genomic data from a 65,000 to 50,000 BP Neandertal female from Vindija, Croatia, to ancient and modern humans, Altai Neandertals, and Denisovans. Ju et al. inserted the human variant of the *ADSL* gene that reduces adenylosuccinate lyase activity in the brain (and elsewhere) into mice and put altered and wild-type mice through a series of tests. The result hailed as proof that a single gene could explain human cognitive superiority (see *Washington Post*, 5 August 2025) was that altered female mice were better at accessing restricted water than wild-type female mice, although altered males outperformed wild-type males on the grip test and there was no difference on seven other tests. Males tied with females on four tests and scored lower on five.

2 Haplogroups are a collection of haplotypes (sets of alleles inherited together on a single chromosome) which share features such as an SNP inherited from a common ancestor. Based on mutation rates it is possible to estimate the number of generations that have passed between samples.

3 Although Denisovan specimens found in Tibet and Siberia suggest specialization for a cold climate, a mandible recently found in Taiwan indicates that Denisovans also inhabited warmer zones (see Tsutaya et al. 2025).

4 The authors note that lower amounts of within-population variation and greater amounts of between-population variation have been found with respect to variants on the non-recombining portion of the Y chromosome, yet these too do not involve functional genes.

5 This conclusion is based on analysis of around 1,300 genomes, 7,000 residues on pottery from 550 European sites covering a time span of 9,000 years, and data on consumption of wild animals, population density, and population instability.

6 Chen and Parmigiani (2007) report a cumulative risk of developing breast cancer at age 70 of 57% for *BRCA1* and 49% for *BRCA2*; of ovarian cancer, 40% for *BRCA1* and 18% for *BRCA2*.

7 The dampening effect of the bacteria on hypoxia was demonstrated in mice, while the presence of *Blautia* is widespread among high-altitude dwellers and increased significantly in human volunteers from low altitude who lived for more than three months at high altitude.

8 The analysis found 57 loci associated with the disorder and was based on around 1 million genomes, about 1% of which came from people who reported stuttering.

9 Graves (2015) cites a different US study that found a much higher proportion of African-origin markers in over 5,000 African American customers of the 23andMe company: 73.2% versus 24% European American and 0.8% indigenous American. The highest proportions of European-origin markers were found in African Americans from Washington state and West Virginia, whereas the lowest were in southeastern states.

10 Starting with two parents who each contribute half the genes through random recombination during fertilization, the proportion of genes falls by half at each generation, while the number of ancestors doubles.

11 With their million or more samples, there is a 90% probability that a database such as the public GEDmatch will identify at least one third-cousin relative of any given individual (see Khan and Mittelman 2018).

## References

1000 Genomes Project Consortium. 2015. A global reference for human genetic variation. *Nature* 526(7571):68–74.

Aarzoo SS, Afzal M. 2005. Gene diversity in some Muslim populations of North India. *Human Biology* 77(3):343–353.

Adhikari K, Mendoza-Revilla J, Sohail A, Fuentes-Guajardo M, Lampert J, et al. 2019. A GWAS in Latin Americans highlights the convergent evolution of lighter skin pigmentation in Eurasia. *Nature Communications* 10(1):358.

All of Us Research Program Genomics Investigators 2024. Genomic data in the All of Us Research Program. *Nature* 627:340–346.

American Society of Human Genetics. 2020. Statement regarding concepts of “good genes” and human genetics. https://www.ashg.org/publications-news/ashg-news/statement-regarding-good-genes-human-genetics/

Anthony DW. 2023. Ancient DNA and migrations. *Journal of Anthropological Archaeology* 70:101508.

Antonio ML, Gao Z, Moots HM, Lucci M, Candilio F, et al. 2019. Ancient Rome. *Science* 366(6466):708–714.

Arciero E, Almarri MA, Mezzavilla M, Xue Y, Hallast P, et al. 2025. Whole-genome sequences provide insights into the formation and adaptation of human populations in the Himalayas. *Current Biology* 35(15):3624–3637.

Ardelean CF, Becerra-Valdivia L, Pedersen MW, Schwenninger JL, Oviatt CG, et al. 2020. Evidence of human occupation in Mexico around the Last Glacial Maximum. *Nature* 584(7819):87–92.

Bastos AP, Krupenye C. 2026. Evidence for representation of pretend objects by Kanzi, a language-trained bonobo. *Science* 391(6785):583–586.
Beall CM. 2014. Adaptation to high altitude. *Annual Review of Anthropology* 43(1):251–272.
Becker NS, Verdu P, Georges M, Duquesnoy P, Froment A, et al. 2013. The role of GHR and IGF1 genes in the genetic determination of African pygmies' short stature. *European Journal of Human Genetics* 21(6):653–658.
Beja-Pereira A, Luikart G, England PR, Bradley DG, Jann OC, et al. 2003. Gene-culture coevolution between cattle milk protein genes and human lactase genes. *Nature Genetics* 35(4):311–313.
Bergström A, McCarthy SA, Hui R, Almarri MA, Ayub Q, et al. 2020. Insights into human genetic variation and population history from 929 diverse genomes. *Science* 367(6484):eaay5012.
Bird KA. 2021. No support for the hereditarian hypothesis of the Black–White achievement gap using polygenic scores and tests for divergent selection. *American Journal of Physical Anthropology* 175(2):465–476.
Bird KA, Carlson J. 2024. Typological thinking in human genomics research contributes to the production and prominence of scientific racism. *Frontiers in Genetics* 15:1345631.
Bird KA, Jackson Jr JP, Winston AS. 2024. Confronting scientific racism in psychology. *American Psychologist* 79(4):497–508.
Bogin B, Smith P, Orden AB, Varela Silva MI, Loucky J. 2002. Rapid change in height and body proportions of Maya American children. *American Journal of Human Biology* 14(6):753–761.
Brown RA, Armelagos GJ. 2001. Apportionment of racial diversity. *Evolutionary Anthropology* 10(1):34–40.
Carter C. 2021. The racial thinking behind ALS diagnosis. *Anthropology News* website, 1 March. https://www.anthropology-news.org/articles/the-racial-thinking-behind-als-diagnosis/
Chen L, Wolf AB, Fu W, Li L, Akey JM. 2020. Identifying and interpreting apparent Neanderthal ancestry in African individuals. *Cell* 180(4):677–687.
Chen S, Parmigiani G. 2007. Meta-analysis of BRCA1 and BRCA2 penetrance. *Journal of Clinical Oncology* 25(11):1329–1333.
Claussnitzer M, Cho JH, Collins R, Cox NJ, Dermitzakis ET, et al. 2020. A brief history of human disease genetics. *Nature* 577(7789):179–189.
Corl A, Bi K, Luke C, Challa AS, Stern AJ, et al. 2018. The genetic basis of adaptation following plastic changes in coloration in a novel environment. *Current Biology* 28(18):2970–2977.
Dipple KM, McCabe ER. 2000. Phenotypes of patients with "simple" Mendelian disorders are complex traits. *The American Journal of Human Genetics* 66(6):1729–1735.
Donovan BM. 2022. Ending genetic essentialism through genetics education. *Human Genetics and Genomics Advances* 3:1–13.
Dougherty MJ. 2009. Closing the gap. *The American Journal of Human Genetics* 85(1):6 12.
Duello TM, Rivedal S, Wickland C, Weller A. 2021. Race and genetics versus 'race' in genetics. *Evolution, Medicine, and Public Health* 9(1):232–245.
Duncan RG, Krishnamoorthy R, Harms U, Haskel-Ittah M, Kampourakis K, et al. 2024. The sociopolitical in human genetics education. *Science* 383(6685):826–828.
Durvasula A, Sankararaman S. 2020. Recovering signals of ghost archaic introgression in African populations. *Science Advances* 6(7):eaax5097.

Etkin NL. 2003. The co-evolution of people, plants, and parasites. *Proceedings of the Nutrition Society* 62(2):311–317.
Evershed RP, Davey Smith G, Roffet-Salque M, Timpson A, Diekmann Y, et al. 2022. Dairying, diseases and the evolution of lactase persistence in Europe. *Nature* 608(7922):336–345.
Fumagalli M, Moltke I, Grarup N, Racimo F, Bjerregaard P, et al. 2015. Greenlandic Inuit show genetic signatures of diet and climate adaptation. *Science* 349(6254):1343–1347.
Graves Jr JL. 2015. Great is their sin. *The Annals of the American Academy of Political and Social Science* 661(1):24–50.
Gupta P, Galimberti M, Liu Y, Beck S, Wingo A, et al. 2024. A genome-wide investigation into the underlying genetic architecture of personality traits and overlap with psychopathology. *Nature Human Behaviour* 8(11):2235–2249.
Haak W, Lazaridis I, Patterson N, Rohland N, Mallick S, et al. 2015. Massive migration from the steppe was a source for Indo-European languages in Europe. *Nature* 522(7555):207–211.
Hage SR. 2024. Language evolution in primates. *Science* 385(6710):713–714.
Hancock AM, Witonsky DB, Alkorta-Aranburu G, Beall CM, Gebremedhin A, et al. 2011. Adaptations to climate-mediated selective pressures in humans. *PLoS Genetics* 7(4):e1001375.
Handley LJL, Manica A, Goudet J, Balloux F. 2007. Going the distance. *Trends in Genetics* 23(9):432–439.
Hatala KG, Roach NT, Behrensmeyer AK, Falkingham PL, Gatesy SM, et al. 2024. Footprint evidence for locomotor diversity and shared habitats among early Pleistocene hominins. *Science* 386(6725):1004–1010.
Hershkovitz I, Weber GW, Quam R, Duval M, Grün R, et al. 2018. The earliest modern humans outside Africa. *Science* 359(6374):456–459.
Holden C, Mace R. 2009. Phylogenetic analysis of the evolution of lactose digestion in adults. *Human Biology* 81(5):597–619.
Hublin JJ, Ben-Ncer A, Bailey SE, Freidline SE, Neubauer S, et al. 2017. New fossils from Jebel Irhoud, Morocco and the pan-African origin of *Homo sapiens*. *Nature* 546(7657):289–292.
Hublin JJ, Lefèvre D, Perini S, Muttoni G, Skinner MM, et al. 2026. Early hominins from Morocco basal to the *Homo sapiens* lineage. *Nature* 649:902–908.
Hublin JJ, Sirakov N, Aldeias V, Bailey S, Bard E, et al. 2020. Initial Upper Palaeolithic *Homo sapiens* from Bacho Kiro Cave, Bulgaria. *Nature* 581(7808):299–302.
Jarvis ED. 2019. Evolution of vocal learning and spoken language. *Science* 366(6461):50–54.
Ju XC, Lee SY, Ågren R, Machado LC, Xing J, et al. 2025. The activity and expression of adenylosuccinate lyase were reduced during modern human evolution, affecting brain and behavior. *Proceedings of the National Academy of Sciences* 122(32):e2508540122.
Källén A. 2025. *The trouble with ancient DNA*. University of Chicago Press.
Karcher NR, Barch DM. 2021. The ABCD study. *Neuropsychopharmacology* 46(1):131–142.
Karnati H, Ying W, Long X, Touma MJ, Smith I, et al. 2024. Minority populations exhibit distinct clinical and genetic features of celiac disease in the United States. *medRxiv*. https://doi.org/10.1101/2024.12.20.24319436
Kassem MA. 2025. Dissecting continental and intra-European genetic structure using chromosome 22 SNPs from the 1000 Genomes Project. *Atlas Journal of Biology*. https://doi.org/10.5147/ajb.260

Khan R, Mittelman D. 2018. Consumer genomics will change your life, whether you get tested or not. *Genome Biology* 19(1):120.
Kimura I, Miyamoto J, Ohue-Kitano R, Watanabe K, Yamada T, et al. 2020. Maternal gut microbiota in pregnancy influences offspring metabolic phenotype in mice. *Science* 367(6481):eaaw8429.
Kirinya M, Linga E, Loki R. 2024. Footprint evidence for locomotor diversity and shared habitats among early Pleistocene hominins. *Science* 386(6725):1004–1010.
Kitada S, Nakamichi R, Kishino H. 2021. Understanding population structure in an evolutionary context. *G3: Genes/Genomes/Genomics* 11(11):jkab316.
Lala KN, Feldman MW. 2024. Genes, culture, and scientific racism. *Proceedings of the National Academy of Sciences* 121(48):e2322874121.
Lee JJ, Wedow R, Okbay A, Kong E, Maghzian O, et al. 2018. Gene discovery and polygenic prediction from a genome-wide association study of educational attainment in 1.1 million individuals. *Nature Genetics* 50(8):1112–1121.
Lewontin RC. 1972. The apportionment of human diversity. *Evolutionary Biology* 6:381–398.
Li L, Comi TJ, Bierman RF, Akey JM. 2024. Recurrent gene flow between Neanderthals and modern humans over the past 200,000 years. *Science* 385(6705): eadi1768.
Lin M, Siford RL, Martin AR, Nakagome S, Möller M, et al. 2018. Rapid evolution of a skin-lightening allele in southern African KhoeSan. *Proceedings of the National Academy of Sciences* 115(52):13324–13329.
López S, Van Dorp L, Hellenthal G. 2015. Human dispersal out of Africa. *Evolutionary Bioinformatics* 11s2:EBO-S33489.
Martin MK. 2023. *The wrong ape for human origins*. Lexington Books.
Martschenko D. 2020. DNA dreams. *Research in Education* 107(1):33–54.
May T. 2018. Sociogenetic risks–ancestry DNA testing, third-party identity, and protection of privacy. *New England Journal of Medicine* 379(5):410–412.
McCartney J. 2023. Exposing the hazards of teaching 19th century genetic science. *Journal of Research in Science Teaching*. https://doi.org/10.1002/tea.21895
McDermott R. 2016. Ethics, epidemiology and the thrifty gene. *Health Psychology* 5:458–466.
McPherron SP, Alemseged Z, Marean CW, Wynn JG, Reed D, et al. 2010. Evidence for stone-tool-assisted consumption of animal tissues before 3.39 million years ago at Dikika, Ethiopia. *Nature* 466(7308):857–860.
Mehrgou A, Akouchekian M. 2016. The importance of BRCA1 and BRCA2 genes mutations in breast cancer development. *Medical Journal of the Islamic Republic of Iran* 30:369.
Mentis AF, Dardiotis E, Katsouni E, Chrousos GP. 2021. From warrior genes to translational solutions. *Translational Psychiatry* 11(130):1–11. https://doi.org/10.1038/s41398-021-01257-2
Novembre J, Stein C, Asgari S, Gonzaga-Jauregui C, Landstrom A, et al. 2022. Addressing the challenges of polygenic scores in human genetic research. *The American Journal of Human Genetics* 109(12):2095–2100.
Pickrell JK, Berisa T, Liu JZ, Ségurel L, Tung JY, et al. 2016. Detection and interpretation of shared genetic influences on 42 human traits. *Nature Genetics* 48(7):709–717.
Plummer TW, Oliver JS, Finestone EM, Ditchfield PW, Bishop LC, et al. 2023. Expanded geographic distribution and dietary strategies of the earliest Oldowan hominins and Paranthropus. *Science* 379(6632):561–566.

Polikowsky HG, Scartozzi AC, Shaw DM, Pruett DG, Chen HH, et al. 2025. Large-scale genome-wide analyses of stuttering. *Nature Genetics* 57:1835–1847.
Potts R, Behrensmeyer AK, Faith JT, Tryon CA, Brooks AS, et al. 2018. Environmental dynamics during the onset of the Middle Stone Age in eastern Africa. *Science* 360(6384):86–90.
Protsiv M, Ley C, Lankester J, Hastie T, Parsonnet J. 2020. Decreasing human body temperature in the United States since the Industrial Revolution. *Elife* 9:e49555.
Prüfer K, De Filippo C, Grote S, Mafessoni F, Korlević P, et al. 2017. A high-coverage Neandertal genome from Vindija Cave in Croatia. *Science* 358(6363):655–658.
Racimo F, Woodbridge J, Fyfe RM, Sikora M, Sjögren KG, et al. 2020. The spatiotemporal spread of human migrations during the European Holocene. *Proceedings of the National Academy of Sciences* 117(16):8989–9000.
Ringbauer H, Huang Y, Akbari A, Mallick S, Olalde I, et al. 2024. Accurate detection of identity-by-descent segments in human ancient DNA. *Nature Genetics* 56(1):143–151.
Roberts NJ, Vogelstein JT, Parmigiani G, Kinzler KW, Vogelstein B, et al. 2012. The predictive capacity of personal genome sequencing. *Science Translational Medicine* 4(133):133ra58.
Rogers AR, Harris NS, Achenbach AA. 2020. Neanderthal-Denisovan ancestors interbred with a distantly related hominin. *Science Advances* 6(8):eaay5483.
Rosenberg NA. 2011. A population-genetic perspective on the similarities and differences among worldwide human populations. *Human Biology* 83(6):659–684.
Sabatello M. 2018. A genomically informed education system? *Journal of Law, Medicine & Ethics* 46(1):130–144.
Sankar P, Kahn J. 2005. BiDil. *Health Affairs* 24(Suppl1):W5–455.
Schmid KM, Lee D, Weindling M, Syed A, Agyemang SL, et al. 2022. Mendelian or multifactorial? *Journal of Microbiology & Biology Education* 23(3):e00093.
Scott EM, Halees A, Itan Y, Spencer EG, He Y, et al. 2016. Characterization of Greater Middle Eastern genetic variation for enhanced disease gene discovery. *Nature Genetics* 48(9):1071–1076.
Slon V, Mafessoni F, Vernot B, De Filippo C, Grote S, et al. 2018. The genome of the offspring of a Neanderthal mother and a Denisovan father. *Nature* 561(7721):113–116.
Stæger FF, Andersen MK, Li Z, Hjerresen JP, He S, et al. 2014. Oxytocin and vasopressin receptor gene variation as a proximate base for inter-and intraspecific behavioral differences in bonobos and chimpanzees. *PLoS One* 9(11):e113364.
Su Q, Zhuang DH, Li YC, Chen Y, Wang XY, et al. 2024. Gut microbiota contributes to high-altitude hypoxia acclimatization of human populations. *Genome Biology* 25(1):232.
Sümer AP, Rougier H, Villalba-Mouco V, Huang Y, Iasi LN, et al. 2025. Earliest modern human genomes constrain timing of Neanderthal admixture. *Nature* 638(8051):711–717.
Tabery J. 2023. *Tyranny of the gene*. Knopf.
Templeton AR. 1998. Human races. *American Anthropologist* 100(3):632–650.
Tsutaya T, Sawafuji R, Taurozzi AJ, Fagernäs Z, Patramanis I, et al. 2025. A male Denisovan mandible from Pleistocene Taiwan. *Science* 388(6743):176–180.
Villanea FA, Peede D, Kaufman EJ, Añorve-Garibay V, Chevy ET, et al. 2025. The MUC19 gene. *Science* 389(6762):eadl0882.
Visscher PM, Yengo L, Cox NJ, Wray NR. 2021. Discovery and implications of polygenicity of common diseases. *Science* 373(6562):1468–1473.

Welker F, Ramos-Madrigal J, Gutenbrunner P, Mackie M, Tiwary S, et al. 2020. The dental proteome of Homo antecessor. *Nature* 580(7802):235–238.
Wray NR, Wijmenga C, Sullivan PF, Yang J, Visscher PM. 2018. Common disease is more complex than implied by the core gene omnigenic model. *Cell* 173(7):1573–1580.
Zhang P, Zhang X, Zhang X, Gao X, Huerta-Sanchez E, et al. 2022. Denisovans and *Homo sapiens* on the Tibetan Plateau. *Trends in Ecology & Evolution* 37(3):257–267.

# 4 Essentialism in science textbooks

## Sex and gender

The last two chapters explored the effects of conventional explanations of genetics on essentialist thinking about human differences, especially between purported biological races. This chapter shifts the focus to simplified and, in many respects, erroneous explanations of fertilization, sex determination, and human origins that foster misperceptions about differences between sexes and genders. By presenting sex determination as a straightforward chromosomal matter, textbooks imply that humans come in two discrete sex categories that are fixed, stable, and internally uniform. Topics that would blur the boundaries are avoided or treated minimally, such as intersex, the homology of reproductive organs, or the circulation of analogous hormones in all humans. An emphasis on biological traits without exploration of the topic of gender allows physical differences to appear to model differences in behavior, aptitudes, interests, and intelligence. Discussions of sex-linked conditions and altered mental functioning in individuals with chromosomal anomalies provide further nudging in this direction.

Explanations of fertilization transpose traditional gender beliefs onto gametes and reproductive systems, naturalizing conventional gender stereotypes and reinforcing a conflated sex/gender binary. Chapters or sections on human origins ground current gender assumptions in alleged productive and reproductive roles across deep time, providing an ultimate explanation for stereotypical aptitudes, interests, and tasks. Evolutionary sequences and trees not only represent humankind as male but also uphold a racial hierarchy whereby White males sit at the apex of evolutionary progress. Together these components of biology textbooks collapse the distinction between sex and gender, suggest that there are only two sexes identical to two genders, and highlight differences while exaggerating within-category uniformity. Textbooks consequently deliver information that is not only incomplete but also out of line with current scientific knowledge.

DOI: 10.4324/9781003639923-4

**Sex and gender in textbooks**

As shown below, science textbooks often use the words "sex" and "gender" interchangeably. This common practice, evident in expressions such as "gender reveal party," reflects the convergence of bio-somatic with bio-behavioral (or neurogenetic) essentialism whereby physical traits taken as evidence of a group's essence are presumed to stem from the same source as mental ones. For ease of reading, the following chapters continue to use the term "genetic essentialism" for both types. The term sex/gender is adopted where sex and gender are treated as if they were the same. Otherwise, sex is used to describe categories defined by chromosomes, reproductive organs, or hormones (male, female, intersex, third sex, trans) whereas gender refers to identities associated with constellations of cognitive, behavioral, and social-role characteristics culturally attributed to and expected of group members (woman, man, nonbinary, trans). Variation in both sets of labels across cultures and historical periods is discussed in the next chapter.

Children's gender-saturated cultural universe includes many kinds of literature which, as the written word, deliver strong messages about gender-appropriate behavior and expectations. As especially authoritative sources of information, textbooks, and particularly science textbooks, help to channel children's interests, attitudes, skills, and knowledge through the stereotypes they incorporate. For instance, repeated exposure to images which equate math and science with males contributes to the deviation of girls away from these fields (Hannover and Kessels 2004). Even a single image embedded in a scientific lesson can affect a student's ability to learn, as shown by an experiment in which 9th- and 10th-grade students without chemistry training were given a chemistry text accompanied by an illustration showing a scientist as either a woman or man (Good et al. 2010). Afterward the students were tested for comprehension. Boys shown a counter-stereotypical image had lower comprehension than those who saw the scientist portrayed as a man, whereas girls shown a stereotypical image demonstrated lower comprehension than girls who saw a scientist depicted as a woman.

Persistent, internationally widespread patterns in children's literature include overrepresentation of males in titles and as central characters, along with traditional gender stereotypes about character traits, social roles, activities, and physical ranges. To illustrate, an analysis of thousands of English-language children's books published throughout the 20th century yielded a 1.6:1 ratio of male to female characters, and nearly double the number of males than females named in book titles.[1] The middle decades of the century showed the greatest gender imbalance (McCabe et al. 2011).

The overrepresentation of males declined from the last decades of the 20th century into the 21st, as shown by an analysis of central characters in thousands of books for readers aged 0–16 available in the US from 1960 to 2020 (Casey et al. 2021).[2] The analysis excluded books with nonbinary central characters. The male-to-female proportion in the last decade under consideration was 1.22:1. Books were divided into six target-age divisions. Overall, there were higher M:F ratios among nonhuman than human characters; books written by men than women; books for younger (0–8) than older children (9+); and non-fiction than fiction books. Nonhuman male characters were overrepresented in fiction, and human male characters were overrepresented in non-fiction. Over time, a decline in male overrepresentation by female authors was concentrated in books with human characters and books for older children (in which female protagonists came to outnumber males), leaving male overrepresentation in books with nonhuman characters, generally for younger readers. Overrepresentation by male authors remained relatively stable among books for older readers, with a reduction only in books for younger readers.

In children's literature, males not only appear as protagonists more often than females, but their traits are presented in opposition to those associated with females. Language and visual images show females paying attention, males seeking it. Female people and animals convey passivity, weakness, emotionality, service, and caregiving. They tend to be pictured in indoor and domestic settings and are shown as housewives or in a limited range of professional roles that tend overwhelmingly to be female-coded, such as teacher or nurse, or fanciful, such as princess or witch. In contrast, male people and animals convey activity, initiative, competitiveness, competence, and leadership. Males more often appear in outdoor and professional settings. Their occupations span the whole arc of human history, from pre-industrial cobblers and cavaliers to scientists, explorers, and astronauts. While female characters are described with adjectives that include negatively valued terms such as vain, chatty, jealous, and evil, the adjectives used in reference to male characters are consistently positive (see Biemmi 2011; Sovič and Hus 2015).

Textbooks across subjects echo the gender associations and numerical imbalances in non-technical literature. Male overrepresentation increases with the target age of readers, as well as the age of the people represented. Girls and boys may even appear in equal numbers and be shown engaged in roughly the same activities, as in German textbooks, whereas images of adults consistently overrepresent men (Moser and Hannover 2014). In Brunei, Elgar (2004) found four times as many images of males as females in secondary-school science textbooks, with an expansion in the gender imbalance as the target age group increased. In addition to appearing more frequently, men are much more often shown working, as in a nearly 2:1

proportion in Australian textbooks (Lee and Collins 2010; see Caldwell and Wilbraham 2018).

Men's work is also represented as more important than women's, with the latter's knowledge and accomplishments devalued relative to men's. For instance, beyond conveying the usual messages about women's passivity and lower status, images in US children's science books on space exploration understate women's participation in the field, depict them engaged in less prestigious activities, and render their contributions inconsequential (Caldwell and Wilbraham 2018). Across subject fields, girls and women are presented in stereotypical ways that greatly misrepresent their actual roles, activities, and spatial range in contemporary society.[3] An analysis of 35 studies on math textbooks published in 16 countries between 2003 and 2024 found that images of professional settings portrayed males in leadership and problem-solving activities, females in supportive and teaching roles (Nandi et al. 2024). Gender-imbalanced presentations were common across countries, including China, India, Qatar, Cameroon, Brazil, Thailand, the Netherlands, and the US, although there was variation in their severity, such as low gender bias in Australian textbooks relative to those published in Singapore and Turkey.

Electronic educational resources likewise overrepresent male contributions to science, as shown by an analysis of banks of English-language materials for primary-school science teachers (Kerkhoven et al. 2016).[4] The frequency of images was highest for boys, followed by girls, men, and, in distant last position, adult women (33.7%, 29.9%, 22.7%, and 13.7%, respectively). Images of children showed boys and girls involved in similar activities, although the moderately higher number of boys resulted in more images of them engaged in most activities, including experimenting and presenting. Neither boys nor girls were shown teaching. In contrast, men were shown much more frequently than women and were greatly overrepresented in images of scientific activity, with the exception of teaching science. Women were much more likely to be shown in other professions than science, whereas for men it was the reverse (46.8% of women are shown in other professions, 30.5% as scientists, and 22.7% as teachers; versus 37.2%, 55%, and 7.7% among men, respectively). About two-thirds of the images of teaching featured women, and only one-third showed men, mirroring the tendency for teachers to be portrayed as women across textbooks in all subjects.

As "bearers of the discipline's status quo," textbooks are slow to incorporate material that challenges accepted beliefs or implicit assumptions, and the lag time in science textbooks is especially long compared to other fields (Metoyer and Rust 2011:180). For instance, evolutionary biology textbooks continue to present an outdated male-oriented view of sexual selection. Males are overrepresented in images and in illustrations of concepts, while women and men are pictured in stereotypical gender roles. Even images of other animal species convey the cultural association between females and reproduction: despite the variability in size differences between males and females across species (see Chapter 5), female animals are presented as smaller and weaker than male animals, and more responsible for parenting (Fuselier et al. 2018).

English-language textbook authors had greatly reduced overtly sexist language by the 1990s, replacing the universal masculine with neutral terms or alternating between male and female pronouns. Nonetheless, textbooks have continued to maintain the male-first convention within sentences and in the ordering of subjects (Campo-Engelstein and Johnson 2014; Lee and Collins 2010). Textbooks written in languages with grammatical gender not only present male before female topics but also convey gender associations through nouns, adjectives, and the universal "man." Regardless of language type, textbooks tend to use the passive voice when discussing females versus the active voice for males.

The stock image of a scientist as a White man continues to predominate in textbooks and the popular imagination throughout the world. For instance, national textbooks in Lebanon most often represent scientists as White men and include few images of non-White scientists (Yacoubian et al. 2017). Teachers and students echo this tendency, as shown by a study in which 7th–9th grade chemistry teachers and their students in Lebanon tended to draw a scientist as a White man working alone in a laboratory, with older children demonstrating a more consistent White-male bias than younger ones (El Takach et al. 2020). In addition to the ethnic message, these images express a Eurocentric vision of science as the individual pursuit of objective truth through laboratory research untainted by personal background, cultural and historical context, or the contributions of others.

In the US, where non-White individuals earn 35% of undergraduate biology degrees and women earn 60%, college textbooks continue to emphasize the contributions of White men. Analysis of seven of the most widely used undergraduate introductory biology textbooks published between 2016 and 2019 reveals that 7% of the 1,107 scientists listed in the indices were non-White and 13% were women (Wood et al. 2020). The proportions increased among named scientists working in successive decades from 1900 on for women and most non-White groups, especially after 2000. While the gender proportions were generally aligned with the gender composition of the scientific workforce (as measured by publications), the complexities involved did not allow the researchers to analyze the correspondence for race/ethnicity. They conclude that it will be many decades before textbooks accurately reflect the gender and race/ethnic make-up of college biology classrooms.

The images in a biology textbook by Sadava et al. (2021) illustrate the patterns described above. The text is a translated English-language book currently used in Italian high schools with a science/technology focus. Nearly everyone in the images, whether photographs or diagrams, is White. Renowned scientists pictured in the book consist of 23 White men and 2 White women. The only Black people in the book are Barack Obama and Henrietta Lacks. Besides 2 large groups of White people in separate photographs, there are 19 non-scientists in the images. All appear to have European ancestry except for one probable Japanese man looking at a monument to the victims at Hiroshima and two Chinese women poultry workers. While 10 are men/boys and 9 are women/girls, 4 of the women are nurses in a single etching shown fretting over a man ill with tuberculosis. Images of scientists show 11 people, of which 10 are White (although two seen in profile have long dark hair) and 1 has Asian features. Scientists include a senior woman and man in conversation at the far end of a laboratory, the woman scratching her head, and three young women working at a bench at the near end. In addition, a woman and a man entomologist are pictured together collecting samples in a tree. The remaining men are one in a lab coat looking intently at a computer, and another wearing safety glasses while examining a gel electrophoresis screen. The other women are a marine biologist shown from the rear and a woman bringing a tray of supplies into an empty lab. Although more women than men are pictured as scientists, only men are shown working alone and only women are shown in lower-status roles.

While pictures of women/girls convey gender stereotypes (the nurses; a mother taking a selfie with her family; a woman shopping), those with men/boys show them standing in for humankind: adult twins wearing identical shirts but tattoos on opposite arms, an infant identified as male by the caption, and two pair of eyes (although there is also one image of eyes in a woman's face). Drawings of generic humans are more often male: nine males, four females. All the female images and four of the male images have light to medium-beige skin tones. The remainder are a purple male, a male in two radiographs, and two diagrams on herd immunity with dozens of identical male symbols in bold colors. All the skin-colored males have short hair, and all the females have long hair. Two of the females are dressed but bare-footed and two are nude, whereas all the males are nude. Three of the female images are paired with a male image, leaving one shown alone, whereas three male images are paired and six are males alone. The one female figure shown alone is clothed and represents humankind in a diagram of the phylogenetic relationships among humans, chimps, gorillas, and orangutans. This image is counterbalanced by a drawing that evokes the traditional concept of woman as companion and admirer of man: the female's head is turned to gaze at the male, who looks straight ahead.

**Images of human bodies in textbooks**

- nude male body treated as standard
- female body represented nude primarily in sections on reproduction
- females soft and curvy, males muscular
- females with long hair, males bald or short-haired
- males in active or dominant poses, females in passive or submissive poses
- bodies almost exclusively young and pale-skinned
- more males than females

*Figure 4.1* Gendered presentation of human bodies in textbook images.

Figure 4.1 summarizes patterns in the images included in science textbooks and discussed further below. In addition to the numerical imbalance in favor of males, images embody stereotypes about physical sex differences and conflate sex with gender through hair and clothing. Images often present more extreme gender stereotypes than the accompanying text, yet their immediacy renders them especially influential (see Moser and Hannover 2014).

Young people are continually exposed to gender stereotypes and male overrepresentation in technical literature as well as fiction books, through both text and images. As seen below, textbook explanations of reproductive biology and human origins are steeped in the same tendencies and breathe life into them in turn.

## Synopsis of textbook explanations of reproductive biology and human evolution

Typically, textbook descriptions of male and male reproductive anatomy feature lengthy explorations of female hormonal cycles, with little mention of hormones in males aside from the effects of puberty on secondary sex characteristics. Textbooks explain the production of gametes through meiosis (see Chapter 2) and the meeting of egg and sperm at fertilization. Far more detail is provided about the sperm's movement from the male reproductive tract to the egg than the egg's movement from the ovary to the fallopian tube. The production and composition of seminal liquid receive considerable attention, whereas cervical mucus is mentioned not at all or described only as an obstacle or impediment to sperm movement and survival. Fertilization is often erroneously described as mechanical penetration of the egg, even though textbooks also mention the chemical events that alter the egg's plasma membrane and prevent the entry of other sperms.

After describing the fusing of the two nuclei, textbooks explain implantation and the zygote's early cell divisions.

These explanations are accompanied by diagrams and electron micrographs that respect a strict sex/gender binary and render the sperm cell the protagonist of fertilization. Couples are drawn or pictured in stereotypical fashion as a woman and a man, coded by long or short hair, small or large body size, softness versus muscular definition, and pink/red and blue tones, respectively. Multiple sperm cell parts are indicated in diagrams (acrosome, nucleus, nucleolus, mitochondria, cytoplasm, flagellum), and the cell is divided into head, neck, and tail. In contrast, the egg's components are limited to the nucleus, nucleolus, cytoplasm, and possibly the zona pellucida, with mitochondria occasionally mentioned. The two gametes may be colorized in pink and blue, and even though some textbooks note that the egg cell is the largest of all cells, images of sperm often take up more space on the page than eggs.

Discussions of sex determination in textbooks focus primarily on the X and Y chromosomes joined during fertilization. They explain the role of the SRY region of the Y chromosome in generating a cascade of events that cause male gonads to form, provided an X-derived anti-testicular factor is also blocked. Female development thereby appears to be an automatic, passive process whereas male development requires an active developmental program. The homology of reproductive parts that share the same embryonic origin is left out of the discussion, as are the commonalities between "male" and "female" hormones, their secretion in bodies of all sexes, and their presence in a vast array of species.

As background, textbooks also contrast asexual against sexual reproduction. They note that the sex which produces gametes carrying two different sex chromosomes is not always the male (for instance, female birds produce eggs with W and Z chromosomes) and may describe how environmental conditions determine whether an individual produces female or male gametes, or both, among many fish, reptile, and other species. Textbooks commonly include a discussion of well-known chromosomal anomalies (missing or extra sex chromosomes) that cause sterility and various anatomical features, and point out behavioral traits such as an increased likelihood of disorders related to social and emotional deficits in men with Klinefelter Syndrome (XXY). These discussions leave out intersex and hormonal variations that can affect reproductive system development out of line with chromosomal sex.

Explanations of sex determination are often accompanied by graphics and karyotypes of the X and Y chromosomes. They include discussions of X-linked recessive genes that cause diseases such as hemophilia and Duchenne muscular dystrophy, as well as conditions such as red-green colorblindness. Rare X-linked dominant genes may also be noted, such as fragile X syndrome and vitamin-D-resistant rickets/hypophosphatemia that can affect females as much as males. The way recessive variants are presented misleadingly suggests that they exclusively affect males whereas females are only carriers. Pedigree charts that illustrate how X-linked genes are transmitted across generations are gendered for humans through circles and squares for female and male, respectively, whereas the same processes in nonhuman species are illustrated with symbols such as chromosomes or red and white eyes in the case of fruit flies. Males are shown to pass the defective allele to all their daughters and no sons; heterozygous females to half their children of both sexes; and homozygous females to all their children. Finally, texts may note that females have an immune-system advantage that reduces the likelihood and severity of infectious diseases but renders them more susceptible to autoimmune disorders (see Chapter 5).

Together, text and visuals present a dichotomous view of sex. Use of the word "gender" interchangeably with "sex," along with discussions of behavioral traits associated with chromosomal anomalies, suggests that physical and health characteristics model differences in cognitive and behavioral traits. Chapters or sections on human origins then provide the ultimate biological cause: the conditions of life over millions of years of evolution, giving rise to separate roles and mental characteristics between females and males.

Discussions of evolution include some of the information presented in the previous chapter on human origins. They place humans in the order Primates and explain the phylogenetic relationships and shared characteristics linking humans and the great apes before outlining the traits that separate them. Textbooks generally name locations, dates, and anatomical features (especially height and cranial capacity) of some of the multiple species that existed (and for varying spans of time co-existed) during the millions of years before and after the emergence of *H. sapiens*. They include descriptions and photographs of famous discoveries in paleontology. Students learn about migrations of early *Homo* species and their settlement across regions outside Africa, as well as the handful of last-remaining species before *H. sapiens* became the only species of *Homo* left. Textbooks often include information on Neandertals and their technical abilities, caregiving, and material culture to indicate their similarity to humans and hypothesize about their ultimate disappearance. While some textbooks note or provide graphics indicating a bush or tree to convey the overlap in the temporal spans of multiple species, many continue to render human evolution a linear progression from an ape-like creature to ever-taller, more hairless, and paler-skinned male exemplars of species leading to humans, the most recent of which carry spears or clubs.

Advances in toolmaking by species of *Homo* feature prominently. In general, textbooks leave the cultural association linking stone tools, hunting, and males intact, despite describing other uses for tools and possibly noting that early humans are more likely to have scavenged than hunted. More commonly, textbooks imply or state that even the earliest humans were hunters and may also suggest that australopithecines managed to kill rather than simply collect and process animal foods. The progressive increase in brain size, as well as the development of human visual and social capacities, is attributed to the cognitive demands of hunting in the grasslands. Despite noting that foraging is a nomadic way of life, textbooks describe male hunters as venturing over vast areas while females took care of children and gathered plant foods close to home, often described as a cave.

Besides the focus on stone tools, narratives about the evolution of *H. sapiens* spotlight parietal (cave and rock) art; the emergence of spoken language; the control of fire; and, protection from predators and the elements. Textbooks describe the effects on food availability of changing climatic conditions, as well as hybridization events with Neandertals and Denisovans and their traces in today's genomes. Finally, there may be a discussion of the Neolithic transition to agriculture and permanent settlements, with farming presented as an improvement over foraging.

In addition to evolutionary sequences, graphics include illustrations depicting scenes of prehistoric life that overrepresent males and depict males and females in stereotypical roles. Reconstructions of the face and bodily form of ancestral species show females with rounded shapes and gentle faces (such as Lucy, shown smiling gently and averting her eyes), whereas males look muscular and wear a grim expression. Drawings depict prehistoric artists as males and imply that paintings of large animals depict hunting scenes. The juxtaposition of these creative renderings against photographs of skeletal remains and footprints makes them appear to be rooted in sure evidence, just as reconstructions based on ancient DNA imply a solid genetic basis, noted in the last chapter. In addition, textbooks foreground sex typing of skeletal remains without explaining the inaccuracies involved in using anatomical measurements, thereby reaffirming a binary view of sex and implying that it has forever determined people's life experience and social roles.

In sum, science textbooks tend to present a limited, essentialist, and in some ways inaccurate vision of dichotomous sex and gender through discussions of reproductive biology and human origins, despite the inclusion of information that suggests sex determination is not as straightforward as implied by the XX/XY model. Textbooks routinely set up oppositions between female and male gametes, reproductive systems, and physical and health characteristics, and avoid discussing overlaps and similarities between them. They fail to distinguish between sex and gender, implying that behavioral traits stem from the same source as physical ones. Discussions of human evolution offer the further suggestion that cognitive and behavioral differences between sex/gender categories ultimately come from an ancient division of labor. Taken together, these components of science textbooks create the impression that the hunter/warrior versus gatherer/nurturer opposition is permanently sedimented in sex-specific genes.

### Gender stereotypes in explanations of reproduction

The use of "sex" and "gender" as interchangeable terms in science textbooks comes with a lack of clear definitions of either term. The word "sex" appears in reference to several different things: individuals with reproductive organs that produce sperms or eggs; sexual desire and behavior; and a variety of physical and physiological characteristics. Sex/gender categories are presented as dichotomous and non-overlapping, such that differences between them are emphasized far more than similarities, and within-group consistency is overstated (see Donovan et al. 2024; Dunk et al. 2024). The conflation of terms suggests the existence of only two forms of sex/gender made up of separate anatomical characteristics, sexual desire and behavior, and aptitudes, interests, and behaviors.

Two interrelated themes run through textbook explanations of reproductive biology: the equation of femininity with reproduction; and the idea that the female body is derivative of the male and therefore less interesting and worthy of direct consideration. These tendencies persist even in anatomy textbooks for graduate and medical students, which uphold a binary concept of sex and gender and deliver gender stereotypes about emotions, roles, and social domains (Parker et al. 2017). Instead of balancing their presentation of human bodies in response to the last half-century's realization that male bodies do not stand in for all, anatomy textbooks continue to include more images of males than females in all chapters except those on reproduction, and some include females only in chapters on reproduction. Female anatomy is described minimally and even erroneously. The paucity of images showing bodies of diverse age, sex, body type, and ethnicity sustains the view of the White, Western, male body as the standard for all humankind (Giacomini et al. 1986; Metoyer and Rust 2011). These tendencies also run through online resources on reproductive anatomy: the amount of information on the female reproductive system is smaller in Wikipedia; there is less discussion of it in the Urban Dictionary; and the information appears more recently in the Oxford English Dictionary (Loughlin 2018).

The lopsided importance assigned to male versus female anatomy extends to descriptions of gametes and reproductive systems. Martin

(1991) shows how 1980s professional literature rendered fertilization a drama involving gendered cells produced by reproductive systems described through value-laden language. New scientific discoveries that contradicted the conventional understanding by which the egg played virtually no role in fertilization were met with the alternative stereotype of women as aggressive, spider-like devourers of men. The discovery of lock-and-key ligands and receptors on the egg and sperm cell membranes brought ill-fitting terminology that upheld gender stereotypes rather than scientific convention. The ZP3 and ZP2 glycoproteins in the egg cell's zona pellucida were given the names "primary sperm receptor" and "secondary sperm receptor" despite their ligandlike structure, whereas sperm-surface receptors became "egg-binding proteins" or "sperm-specific proteins." More recently, the paired proteins on the egg and sperm plasma membranes have been named Juno, for the Roman goddess of fertility, and IZUMO, for a Japanese shrine honoring the Shinto god of marriage. Celebrating the discovery of Juno, a scientific paper proclaims, "IZUMO1 finally finds his mate" (Anifandis et al. 2014:12983).

Stereotypical renderings of egg and sperm, the respective reproductive systems that produce them, and their meeting in the female reproductive tract continue to run through biology textbooks from the middle school through medical school level (Campo-Engelstein and Johnson 2014; Loughlin 2018; Metoyer and Rust 2011; Whitaker and Baccolini 2024). Textbooks provide more information, more correct information, and longer explanations of male than female gametes and reproductive systems. Males are presented as the standard to which females are compared: masculine words come before feminine ones within sentences; male topics appear before female and are described with greater detail and accuracy; and, embryonic development is presented as an active process in males but a passive outcome in females. The generic humans in illustrations of non-reproductive body parts or systems are overwhelmingly male. Female topics are explained with passive language, male topics with active verbs. Reproductive systems appear to have opposed qualities: a cyclical, obscure, complex one that tends toward degeneration versus a linear, straightforward, endlessly productive one. Accordingly, the vagina is portrayed as a mere location, the penis as a functional organ. Where homology is noted, it is only in the direction of male to female: for instance, the clitoris is rendered a downgraded version of the penis.

Eggs are described as immobile, demure, featureless, and few; sperms as energetic, purposeful multitudes with numerous internal and external structures. Fertilization is rendered an exciting adventure, romantic quest, or sacred mission in which resolute sperms overcome dangers and obstacles on the way to a serene, mystical egg that drifts around and is surrounded by "vestments." The female gamete and body are acted upon by their male counterparts, and the delivery of male chromosomes to the egg is depicted as a violent, mechanical process of penetration. Likewise, the joining of chromosomes inside the egg is erroneously attributed to the sperm cell's nucleus moving through the egg cytoplasm and bursting through its nuclear membrane.

These ways of presenting the material are unnecessary and contrary to scientific knowledge, as shown by an analysis of fertilization narratives in 16 frequently used English-language biology textbooks divided equally among the middle school, high school, university, and medical school levels (Campo-Engelstein and Johnson 2014).[5] Unbalanced presentations included detailed descriptions of sperm transport versus little or nothing about egg transport after release from the ovary, no information on the way the cervix and its mucus protect sperms and favor their movement, and nothing about the mechanical help provided by muscular contractions, peristalsis, and cilia. Entire sections on male topics such as sexuality, sexual arousal, or infertility had no corresponding section about females. Alternatively, some books discussed the same topic in relation to females and males but left out information about the former, such as a case in which a book explained the reason for increased blood flow during sexual arousal only with respect to males. Campo-Engelstein and Johnson explain that textbooks which leave out female orgasm and the clitoris reinforce the equation of femininity with reproduction, heteronormative sexuality, and sexual passivity, and further assign greater agency to males and sperms in the fertilization process. The lack of a separate section on sexuality, and of any discussion of non-reproductive sexual behavior, conveys scientifically inaccurate heteronormative assumptions about reproduction and sexual preference and behavior.

Further contrasts included dull language in reference to eggs and the female reproductive tract versus enthusiastic, admiring language concerning sperms and the male reproductive tract. Alkaline mucus produced by the male reproductive tract was described positively as protecting sperms from the harsh acidic environment of the vagina, while the protective effect of cervical mucus was left out. Fertilization was routinely presented from the sperm's point of view, which was concretized by exercises that asked students to map out its route but not the egg's. Gendered similes masculinized the sperm, as in the equation of the acrosome with a motorcycle helmet, or the sperm's abandoned plasma membrane with the skin shed by a snake. Words having to do with journeys, tests of skill, organized assaults, and death and survival anthropomorphized sperm cells, while descriptions of egg-sperm interactions as assault or penetration communicated an aggressive vision that is at odds with scientific understanding and reinforces the symbolic opposition between male strength and female weakness.

An analysis of an educational animated film and nine textbooks for Italian middle schoolers corroborates these findings (Whitaker and Baccolini 2024).[6] The widely viewed film openly anthropomorphized cells, explicitly likening eggs and sperms to women and men. Sperms had faces and wielded weapons; maternal and paternal chromosomes with gendered features danced as couples; and embryonic cells dressed in human garb such as cowboy outfits were directed by a bearded wise man in a control tower. In contrast, the textbooks employed all the indirect means listed above. They reserved the language of awe and wonder for sperm cells, and, in both text and illustrations, celebrated the "only one" sperm that victoriously fertilized the inert egg thanks to its especial strength and valor. Descriptions and drawings of fertilization by three classes of middle schoolers illustrate the impact on student understanding. While some students used neutral language and drew a complex egg membrane rather than a smooth line, in general they tended to follow the active/passive voice tradition, to know much more about sperms than eggs, and to draw sperms vigorously swimming in a straight line and pushing against or entering the egg by force. Some sperms had eyes.

Despite their advanced level, gynecology textbooks also emphasize male topics, convey gender stereotypes, and present incomplete or erroneous information about female reproductive biology, as shown by an analysis of 10 textbooks assigned in US medical schools (Metoyer and Rust 2011).[7] Words, phrases, and sentences were coded for active and passive verbs (or a combination) as well as activity/importance (function, independence, mobility, dominance, strength) and passivity/unimportance (weakness, fragility, immobility, inertness). Overall, the assumption of natural female passivity came through in a vision of the female body as a mere conduit for male-generated young; notions about female brains being smaller and more rudimentary than male brains; and a generalized downplaying of the activity and importance of eggs and cervical mucus, with exaggeration of the activity and importance of sperms and semen.

Although the textbooks avoided the romantic conquest or heroic quest narrative and used some egalitarian language such as "unite" or "meet" when discussing fertilization, traditional gender stereotypes nonetheless shaped the way they presented the material. Female anatomy was described in terms of multiple fragmented parts, whereas the male body appeared active, cohesive, and organized. Sperms were mentioned far more often than eggs: 100 versus 38 instances. Eggs were described predominantly in passive terms, very rarely with a mixture of active and passive voice, and never in completely active language. Active language was the preponderant type used for sperms and often conveyed penetration; active and passive terms never appeared together; and passive terms were very rare. Compared to descriptions of the egg, mutual terms such as "unite" or "meet" – always a minority – were far less frequent.

In line with these patterns, the textbooks generally omitted information about the activity of the cervix and its mucus while lending agency and importance to semen. The cervix was almost always described as an inert, passive location or conduit (85% of cases). In some cases, it appeared fragile or vulnerable to disease (11%); in others, changes to its softness or dilation were noted (4%). There were no counterbalancing statements about its resilience, elasticity, or role in protecting, storing, filtering, and moving sperm cells. Every mention of cervical mucus left out these same properties, describing it mainly in relation to ovulation and secondarily as an obstacle to sperm movement. In several discussions of sperm movement, there was no mention of cervical mucus at all. The power and potency of sperms and semen came through both in the depiction of sperms reaching the egg by their own efforts once given a boost by semen, and the admonition that, to prevent pregnancy, semen must not be spilled anywhere near the vagina.

A comparison between textbooks and contemporary medical and scientific journals verified the tendency for textbooks to omit relevant scientific findings that contradict conventional narratives, such as the knowledge that changes to cervical mucus in the days surrounding ovulation increase its ability to transport sperms (Metoyer and Rust 2011). At the same time, scientific literature continues to describe fertilization through tropes about active, courageous, self-propelled sperms overcoming impediments and challenges on their way to the egg, as in the statement that sperms "swim towards" the uterus and oviduct, "in an attempt to make it through the mucus," while uterine contractions "assist" sperm cells in their "journey" (Georgadaki et al. 2016:979).

Figure 4.2 summarizes the narrative techniques by which textbooks and scientific literature on human reproductive biology express and reinforce

**Descriptive devices in textbook explanations of reproductive biology**

- more comprehensive, accurate information about male reproductive system than female
- males as the standard to which females are compared: masculine words before feminine ones within sentences; male topics before female
- passive voice for female, active voice for male
- opposed reproductive systems: female as cyclical, obscure, complex, tending toward degeneration; male as linear, straightforward, endlessly productive
- homology of reproductive structures often omitted

*Figure 4.2* Gendered descriptions of human reproductive biology in science textbooks.

gender essentialism through descriptions that assign agency, activity, and strength to male bodies, secretions, and cells, while describing female equivalents as passive, weak, acted-upon objects. The persistent habit of granting personhood to sperms and eggs naturalizes cultural beliefs about gender and misconstrues scientific understanding. Similar essentializing biases run through explanations of sex determination and human origins, as seen below.

## Sex determination and differences

Biology textbooks convey genetic essentialism through both language and content related to sex determination and differences, as shown by an analysis of six textbooks used in one or more of the four most populous US states and estimated to be assigned in two-thirds of biology classrooms nationwide (Donovan et al. 2024).[8] Despite the dissimilar political landscapes of California, New York, Florida, and Texas, the study found no difference between them in the frequency of the variables under consideration, suggesting that an essentialist vision runs across the political spectrum. Of 216 chapters, 10 discussed genetics and sex determination, generally portraying sex/gender categories as binary and discontinuous rather than emphasizing the variation within or overlap between them. The terms "sex" and "gender" were used interchangeably and not defined. None of the 362 paragraphs which used the words "sex" or "gender" differentiated between them.

Paragraphs were coded for the four elements of gender essentialism: uniformity, categorical difference, internal causal explanation, and between-group variability. Only 3% of paragraphs described sex/gender categories as continuously variable; 10% included variation within a category, such as blue- versus brown-eyed women; and 12% described members of the same group as uniform. While 11% of paragraphs described similarities across groups, 16% presented sex/gender categories as discrete and non-overlapping – a difference that is not statistically significant but nevertheless signals essentialism given that no paragraphs should have described groups as discontinuous, and none discussed intersex.

Regarding causal explanations for sex/gender differences, 1% of paragraphs referred to external forces, compared to 12% of paragraphs that discussed internal factors. Internal-factor explanations were divided equally between single and multiple causes (gene or hormone versus several alleles). The texts did not explore the complexities of gene-environment interactions, and, moreover, used essentialist descriptions equally with respect to reproductive traits and behavioral/cognitive ones. Consequently, they projected the sense that variation within and between sex/gender groups is due entirely or almost entirely to internal genetic and hormonal causes.

To illustrate how explanations could be changed to present more accurate knowledge, Donovan et al. (2024) provide an alternative reading to an excerpt from a widely used biology textbook. The original states that,

> With X-linked traits, males are more likely to be affected. Males only have one copy of the X chromosome. This means that if they have just one copy of the allele for a disorder, they are affected by that disorder. Females, on the other hand, have two copies of the X chromosome. If they have one copy of the allele for the disorder, females are only carriers of the disorder, not affected by it.

The paragraph sets up discontinuous groups; suggests uniformity within groups; attributes differences to a single, internal cause; and describes variation within groups as falling into discrete types. The revised paragraph points to the existence of sex chromosome combinations other than XX or XY, multiple causation of traits, and continuous variation within and between groups:

> Because of selective X inactivation (the process in which one X chromosome in females is silenced), human XX females can carry X-linked disorders without showing any symptoms. However, several other biological mechanisms can still cause female carriers to show symptoms of X-linked disorders. The percentage of female carriers who show symptoms varies dramatically across disorders for this reason, but the symptoms are generally mild. By contrast, X-linked disorders tend to affect a high percentage of XY males (there is high penetrance) because they do not have a 'spare' X chromosome with a functional copy of the affected gene. In human males, the severity of symptoms of X-linked disorders varies.

A Finnish study illustrates the patterns outlined above while also exploring change over time. Aivelo et al. (2024) analyzed 12 national textbooks used in middle and high school biology classes and published between 1999 and 2019. Finnish textbooks are revised according to national curricular standards that change at roughly 10-year intervals. A particularly consequential modification in 2001 instituted health studies class, which resulted in the offloading of many topics related to sexuality and gender, along with the mandated inclusion of minimal information about intersex and nonbinary identities.

The researchers translate the Finnish word, *sukupuoli*, as sex/gender given that the language does not include separate words for sex and gender. They coded instances throughout the entire books in which sex/gender terms were used, dividing them between comparative uses and ones in which no contrast was made. Given the lack of grammatical gender in Finnish, the words were ones that referred directly to people, such as woman, female, and feminine. In addition, the team counted figures of people in the illustrations; no animals were presented as sexed/gendered.

A dichotomous presentation ran through the text of all the books. All sex/gender contrasts reflected a strict binary and were not accompanied by descriptions of variation within a sex/gender category, although in some cases the statements noted that the differences were between averages. The only exception was text boxes, where intersex conditions and nonbinary gender identities were discussed in all the more recent textbooks. Recent books also included illustrations showing romantic feelings between people of the same sex/gender, although they were much less numerous than photographs of one girl and one boy together.

The contrasts between sex/gender categories fell into two groups: biological and non-biological. The first concerned chromosomes, reproductive anatomy, hormones, body size, hair, body fat distribution, and sexual and reproductive behavior.[9] Biological explanations also extended to academic achievement along with academic and professional inclinations. Differences attributed to non-biological causes included increased mortality in men who use alcohol and tobacco, increased difficulty achieving orgasm in women, and both better dietary habits and greater frequency of eating disorders among women due to psychological and cultural factors that cause them to worry more about their weight than men.

The most frequent explanation of sex/gender differences was hormones, which some books offered as the sole reason for physical and behavioral differences. Examples include the assertion that women lose more bone mass because of declining estrogen production, and the statement that boys have more trouble succeeding at school because of the androgen-induced sex drive that peaks during the same age at which they are expected to study. Hormones were also central to discussions of nonbinary identities and sexual orientations. Less frequently offered explanations involved psychological factors and evolutionary/genetic causes, such as the notion that displays of physical attraction between boys and girls are human universals rooted in sex differences among animals; the hackneyed idea that reproduction is weightier for females due to their producing fewer gametes; and the suggestion that the latter difference causes disparate sexual behavior.

In line with the emphasis on reproduction and sexual development, the analysis yielded more feminine than masculine words, with the word "mother" accounting for almost half of the former. In contrast, proper names were more often masculine, in reference to scientists, celebrities, and male subjects in study exercises. The textbooks embedded gender stereotypes in the narrative and illustrations, such as sociality, emotivity, and diligence in women, or competitiveness, aggression, and risk-taking in men. They thereby implied that "female" and "male" hormones determine gender-stereotypical traits. The authors summarize these traits with an original drawing in which a wide-hipped, soft-bodied woman is shown holding a baby, her internal reproductive parts highlighted. A thought bubble above her head contains emojis beside faces of people whose interrelationships are indicated by lines connecting them. The man, whose internal reproductive parts are absent, is muscular and taller than the woman. He has a cigarette in one hand, a bottle in the other. His thought bubble contains a rod and washer above two mathematical equations.

Regarding change over time, the study shows that the narrowing of the focus to sexual development and reproduction due to the creation of health studies class brought a decline in the average number of sex/gender contrasts from one decade to the next, due largely to a decline in behavioral contrasts. All types of explanation declined, with hormonal explanations declining the most. While the number of gendered words declined, the proportions remained balanced toward feminine words except in the case of proper names. Throughout, textbooks presented the male role in reproduction as ending with fertilization. No exploration of fatherhood was included, nor any alternative version of parenthood.

The sex/gender contrasts in newer textbooks were presented in less declarative terms, such as describing puberty in relation to reproductive organs rather than a transition from girl to woman and boy to man. However, they continued to describe hormones as female and male, offsetting a trend toward providing students more accurate information about hormones as hormonal explanations for sex/gender differences declined. Illustrations included counter-stereotypes such as girls and boys engaged in sports coded as opposite gender, yet newer textbooks were less likely to discuss non-heteronormative sexuality than earlier books (although the earlier books did so in an othering way, such as by prompting students to reflect on the legal right to marry for same-sex couples). The rarely discussed topic of non-reproductive sexual behavior (including contraceptive use) disappeared from newer textbooks, which also avoided discussing nonbinary sex/gender identities in the main text.

These shifts give biology textbooks the appearance of presenting only scientific certainties about human reproductive anatomy and behavior. To raise students' awareness of the sociopolitical context in which science is embedded (see Chapter 2), the researchers suggest discussing the disproportionate scientific interest in the penis compared to the vagina and the dearth of attention to women's bodies due to the alleged complexity caused by hormonal cycles. They argue that, instead of presenting human biology as the basis for teaching about animal physiology, Finnish textbooks should bring information on other species into their explanations of human reproduction. This would make it easier for students to understand the variability in sex and parenting modes among living things, help them appreciate the diversity among humans, and reduce the force with which textbook presentations naturalize cultural beliefs about sexuality and parenting. The authors point out that maintaining a hegemonic vision of binary sex/gender renders school a site of oppression for many students. Direct discussion and illustrations of sex/gender diversity in textbooks, instead of silence, would better represent the variability in the world. Moreover, exploring this diversity would not only be engaging to students during a phase of life in which they are curious about their own biology and subjectivity but also speak to a central concern in biology.

## Student perceptions of sex, gender, and innate ability

The conflation of sex and gender in biology textbooks, together with the presentation of hormonal and evolutionary explanations of human behavior, effaces the cultural content incorporated into the material. Students tend not to notice gender biases embedded into their textbooks, although they can be made aware of them and can develop a critical perspective (see Campo-Engelstein and Johnson 2014). Without prompting, they are better able to distinguish between sex and gender in the case of an infant than a child or adult, as shown by an experiment in which students in three university biology courses were asked to edit two sample questions developed from practice problems in six introductory biology textbooks (Dunk et al. 2024).[10] The books followed the pattern described above whereby sex and gender are not defined or distinguished from one another. Gender was not defined in any text or glossary. One book used the term "gender determination" in place of "sex determination" in the index and text. In another, there was an entry for sex in the glossary which referred the reader to the entry on sexual reproduction. Practice problems routinely used "woman" and "man" in place of "female" and "male" (along with juvenile equivalents), as well as heteronormative expressions involving married couples and their children.

Students were tasked with making one of the two questions more accurate, defined as "correct; precise; using language that applies to all people." Both questions concerned sex determination in human infants and included three instances of gendered language (baby's gender, gender of the baby, woman and her husband, mom, dad, boy). In all three courses, instructors distinguished between sex and gender during lecture. About three-quarters or more of the students in each of the classes identified the terms related to the baby's gender as inaccurate in the question they were given to correct (72–90% regarding "gender of the baby" and 75–77% regarding "baby's gender"). Students in the course on human and animal physiology for biology majors and the introductory course on the biology of sex for non-majors made more infant gender-to-sex edits than students in the introductory biology course for non-majors. High-performing students (over 80% score on the final exam) also made more such edits, although the difference was modest. In addition to substituting "sex" for infant "gender," students made changes such as replacing "woman" or "man" with a sex or chromosomal designation or changing "woman and her husband" to "couple." These edits were less frequent and occurred at similar rates across courses (in the 30–60% range, combining types of edits, for woman/husband, mom/dad, boy, and dad). The study shows that sex/gender conflation is less visible with respect to children and adults, that it persists despite teaching on the difference between the terms, and that more specialized biology training and higher academic performance are associated with greater awareness.

To verify that conventional explanations of sex differences in textbooks cause students to overestimate cognitive/behavioral similarities within groups and distinctions between them, Donovan et al. (2019) randomly assigned one of three texts to 8th–10th-grade students: two were mainstream biology texts on sex differences, one of which discussed plants, the other, humans; the third refuted genetic essentialism by exploring sociocultural explanations for gender differences in Ph.D. attainment in science fields. The first two texts presented sexes as uniform, discrete, and determined by chromosomes. The refutational text was the only one to use the words "woman," "man," and "gender." Afterward, students assessed the amount of difference in bodies, brains, intelligence, science aptitude, and Ph.D. completion and estimated the degree to which genes, the social environment, and personal choices accounted for it. Compared to the pre-test results, students who read the first two texts expressed increased genetic attributions on the post-test instrument, with a slightly higher level of increase after reading the text on human as opposed to plant sex differences. That is, the first two groups interpreted gender differences in science ability through concepts about the genetics of sex. In contrast, genetic essentialism declined among students who read the refutational text, although not as much as expected, possibly due to the potential of refutational texts to produce the opposite outcome (see Chapter 2).

In girls but not boys, the boost in genetic essentialism upon reading the first two texts was accompanied by increased belief that science ability is innate. Students who read the refutational text demonstrated a modest decline in belief in the innateness of science ability; however, segregating the responses between boys and girls revealed that agreement with the innate ability idea did not change among girls. Finally, in girls but not boys, reading the first two texts brought reduced future interest in studying science and technology. Both the pre-test and post-test responses revealed that, for girls, genetic essentialism was positively associated with the innate ability idea, and both were negatively associated with future interest in STEM. Among boys, these three variables were unrelated to one another both before and after the intervention. Consequently, the overall increase in belief that science ability is innate was due primarily to an amplification of girls' preconceptions upon reading the conventional texts.

A follow-up analysis of the students' narrative responses to the experiment explored the use of the terms, "sex" and "gender" (Stuhlsatz et al. 2020). The group that read the refutational text was more likely to distinguish between sex and gender than the two groups that read about the genetics of sex. Those groups used the terms interchangeably, with more of this tendency in the group that read about humans than the one that read about plants.

The authors suggest that girls may be motivated either to reject the innate ability idea to shield themselves against negative gender stereotypes in science courses; or, alternatively, to embrace it as explanation for their low sense of belonging. Boys, in contrast, benefit from the innate ability concept. They automatically fit in thanks to gender stereotypes and numerical disparities in science fields, both of which favor future interest in STEM (see Chapters 1 and 5).

The above results indicate that conventional science education has an especially negative effect on girls and women, who also are more likely to self-stereotype when triggered by genetic essentialism. Belief in genetically determined abilities begets fatalism, whereas the belief that traits are malleable brings the possibility of improvement through effort. Donovan et al. (2019) explain that, while conventional approaches encourage girls to believe they are destined to have poor science ability, teaching about the malleability of aptitudes increases girls' interest in continuing to study male-associated subjects, as has been shown with respect to math competence. Consequently, teaching how science ability may be improved through effort should increase retention rates of girls in science fields. The studies also affirm that biology instruction should emphasize the multifactorial causes of phenotypic traits, explain that genetic factors related to sex are separate from culturally constructed gender, and examine the social roots of unequal gender participation in STEM. In this way, textbooks can break the chain of repeated messages about sex differences in science and society that reinforce and entrench essentialist biases.

## Human origins and evolution

Like explanations of sex determination, textbook presentations of human evolution foreground and overrepresent males except where reproduction is concerned. As a result, they convey an erroneous male-oriented view of prehistoric life and the achievements of human ancestors. Females are represented primarily as mothers, reinforcing traditional gender beliefs about reproductive versus productive roles in the deep past. At the same time, visual images project an interpretation of evolutionary progress as culminating in White civilization.

A European BIOHEAD-Citizen project (see Chapter 1) analyzed the topic of evolution in 29 science textbooks and 1 history book from 12 countries representing diverse cultural and religious traditions inside and outside Europe: Italy, France, Germany, Finland, Senegal, Cyprus, Lebanon, Estonia, Lithuania, Romania, Hungary, and Poland (Quessada et al. 2008).[11] Six other countries could not be included because their textbooks lacked chapters on evolution (Tunisia, Morocco, Algeria, Mozambique, Malta, Portugal).

The images overwhelmingly pictured adult White men, with few women and only rare images of other ethnicities. There were no images of women in 30% of the textbooks. These were from Senegal, Lebanon, and Romania. At the opposite extreme, in Finnish books the numbers were balanced. At the higher end of the scale of inclusion of women were books from Germany, Poland, Hungary, and Estonia. Proportions varied depending on publishers in Cyprus, France, Italy, and Lithuania. Of the 606 depictions of *Homo* in 203 images, there was one female for every four males: 11% females, 44% males. Nine percent were children, while age could not be determined in another 12% of cases. Nearly half were skulls and skeletons that did not indicate sex (45%). Females/women primarily appeared in representations of prehistory and very rarely as scientists, nearly all of whom were males/men.[12]

Ethnic type could not be distinguished in more than half of the images (58%), such as those of skeletons, skulls, or shadows. Among images in which ethnic type could be identified, 90% depicted White people. The remaining 10% showed more than one ethnic type: 4% showed two, generally light- and dark-skinned; 3% showed three; and 3% showed four or more. These books were from Germany, Poland, France, Hungary, and Finland. A typical image reproduced in the report (from an Italian book) shows a male bringing a deer slung over his shoulders to a female who sits at the entrance to a cave, sewing, and surrounded by pots. Both have pale skin and blond hair. Another image (from France) compares australopithecine and modern human females in two poses: the australopithecine is dark skinned, the human light skinned.

There were 62 timelines and evolutionary trees in the textbooks, none of which appeared in books from Finland, Hungary, or Poland. Most of these images depicted a linear process of human evolution culminating in a White male. Of 67 humans in the images, only 6% were female, and all of them were pictured beside a male (Italy, France, Lithuania) or two males

(Germany). Only two images (3%) contained non-White (African and Asian) representatives of *H. sapiens*. Moreover, clothing and lifestyle cues in the images were uniformly European. More than half of the images of males showed them nude (59%). The rest were made up of 18% wearing a business suit (including a White man in a book from Senegal) and 23% with casual Western clothing. A man jogging, a boy on monkey bars, and images of men carrying things – whether businessmen with attaché cases or nude men holding modern tools – further convey Europeanness along with traditional gender stereotypes.

The authors suggest several reasons for the widespread depiction of human ancestors as White males, such as the use of translated English-language texts. Evolutionary timelines and trees that show nude White males may reflect the artistic tradition of representing humankind as Adam. Other factors include religious and cultural beliefs that refute the common ancestry of humans and nonhuman primates, as well as the idea of evolutionary progress crowned by European civilization, which casts non-European peoples as primitives.

In another study as part of the same project, Quessada and Clément (2011) analyzed 50 evolutionary timelines and trees that included an image of *H. sapiens* in 18 science textbooks from 9 countries: Lebanon, Portugal, Romania, Senegal, Tunisia, Cyprus, Estonia, France, and Italy.[13] In no case was *H. sapiens* represented solely by a female. Two images showed a couple (one from France, one from Italy). Aside from 11 images with no identifiable sex (1 shadow, 1 brain, 9 skulls), the remaining 37 were lone White males/men: 23 nude males; 9 men wearing a suit; and 5 men dressed in other clothing such as slacks and a shirt. No image represented a non-White ethnicity. Bush-like images were very rare, whereas linear, goal-oriented schemas were common. The most typical was a series of primates and ancestral species leading to a nude bearded male.

The analysis of Italian middle school science textbooks mentioned earlier (Whitaker and Baccolini 2024) included an analysis of evolutionary narratives. In line with the above studies, males were vastly overrepresented. Evolutionary progressions of hominid species exclusively depicted males, often with weapons in the hands of more recent species. One such progression had a large X across the lineup of species exemplars, the purpose of which was to point out that species relationships should be represented as a tree or bush, rather than the error of representing all species in male form. In addition, there were temporal progressions from earlier life forms such as dinosaurs to humans that typically culminated in White men depicted in modern dress and activities, such as a businessman or astronaut. In an exceptional image, humankind took the form of two women who, however, were drawn according to traditional gender stereotypes: each had long hair, and one kneaded bread while seated as the other stood with a load of grain in her arms. Visual reconstructions of prehistoric life represented males in active, productive poses and roles, whereas females tended to be shown close to home, involved in domestic tasks, and caring for small children. Together with the use of the Italian words for man (*uomo*) and mankind (*uomini*), images and narratives about art, technology, and language delivered the unequivocal message that evolutionary advancements were the fruit of male genius.

The tendency for images to exalt the roles and activities of males in human evolution is common across visual formats, including films, documentaries, museum exhibits, online publications, literature, and even professional scientific publications (see Wiber 2006). A couple of graphics from *Science* magazine illustrate the way gender stereotypes infuse representations of the past. The first, in an article on cross-cultural variability in gut bacteria, presents a decline in the prevalence of cellulose-digesting organisms from nonhuman primates to humans in industrialized societies (Moraïs et al. 2024). There are five block figures in black ink, the first three in profile: a monkey; a "Paleolithic" man crouched and bearing a club; and, a bow-toting man poised to shoot an arrow who represents "Hunter-gatherers." Next, an unlabeled person, shown from the rear, holds a spade and wears a hat and long cloak that could suggest a woman. Finally, man in a business suit and holding a cell phone is shown from the front, with white cutouts for his collar and the cuff of one sleeve.

A second image in *Science* accompanies an article on using ancient DNA to trace genetic networks in Paleolithic humans and Neandertals (Bergström and Tyler-Smith 2017). A group of Neandertals is pictured in a cave. In the foreground, on the left side of the illustration, are a seated female with two small children. The larger child rests its head against her shoulder and presses a hand against its ear, while the other stares blankly ahead. The female stares upward, concerned. Beginning at the midline of the artwork is an older female wearing animal pelts that expose one breast. Eyes downcast, an expression of worry and weariness on her face, she kneels before the fire, on the edge of which rests a large tusk that explains the injured, bearded male at the far right of the scene. Illuminated by firelight, he lies supine on the ground dressed in animal pelts up to his breast. Watching over him is a grown child with long hair wearing a sleeveless shift.

Images such as these reflect the symbolic association linking males to hunting and females to childrearing and caregiving. Sex and gender are conflated through tropes including long or short hair, gender-associated clothing, and stereotypical roles, mirroring the use of the two words as equivalents. Descriptions and visual reconstructions of prehistory consequently reinforce the essentialism that runs through explanations of reproductive biology and sex determination and supply it an ultimate cause: genes already embedded in prehistoric genomes. These sources meanwhile lend support to race/ethnic essentialism by painting human evolution in stages with White populations in top position.

Despite its central importance to biological science and direct connection to genetic inheritance, evolution is often excluded from middle and high school education and is not always a required course for a degree in biology or medicine, thanks in large part to creationism (Graves 2015; Plutzer et al. 2020). Polls in the US show that half or more of the population believes that both creationism and evolution should be taught in public schools (see Pobiner 2016).[14] This proportion indicates widespread denial of the epistemological differences between science and religion along with belief in the divine creation of humans in their present form, including sex and race subtypes (see Chapter 1).

Creationist belief is inversely related to years of university education, as shown by questionnaires distributed to primary and secondary school biology and national-language teachers in 14 countries as part of the "BIOHEAD-Citizen" project mentioned above (Quessada and Clément 2011).[15] Countries were chosen to represent a variety of geographical, historical, and sociocultural characteristics, including European versus non-European, different regions within Europe, various levels of religious observance, and majority faith (Catholic, Protestant, Orthodox, Muslim). Only 19% of teachers with more than five years of university education chose the most creationist of four statements, which ranged from completely divine creation to strictly evolutionary processes as the explanation for human origins. The same statement was chosen by 36% of teachers with three or four years of university education and 53% of teachers with two or fewer years. The project divided teachers into two groups: biologist (primary school teachers with a biology diploma and secondary school biology teachers), and non-biologist (primary school teachers without a biology diploma and secondary school national-language teachers). Biologist teachers were more likely to espouse evolutionary beliefs in nine countries where evolution was taught (Romania, Cyprus, Lebanon, Estonia, Hungary, Finland, Italy, France, Portugal). There was little difference between the two groups in four countries where it was not taught (Algeria, Burkina Faso, Morocco, Tunisia) and one country (Senegal) where it was taught.[16] In other words, biologist teachers generally expressed weaker creationist belief, except in countries where evolution was not taught.

Creationist beliefs were much weaker among teachers in Europe than elsewhere. Fewer than 40% of teachers in France, Estonia, Finland, Italy, Portugal, and Hungary chose either of the two most creationist statements (creation only; creation plus evolution), compared to 60–98% of teachers in Cyprus, Burkina Faso, Romania, Tunisia, Lebanon, Senegal, Morocco, and Algeria, in ascending order.[17] Among Christian respondents, agreement with the most extreme creationist statement varied considerably: under 10% of teachers in France, Estonia, Finland, and Portugal; 10–20% in Italy and Hungary; 20–40% in Cyprus, Burkina Faso, and Senegal; and, 40–60% in Romania and Lebanon.

These proportions indicate that it is not just religious beliefs but also sociocultural context that shapes how teachers interpret human variation. They demonstrate that more years of university education, whether in science or other fields, reduce creationist belief and consequently better equip teachers to communicate accurate information about human variation.

## Anti-essentialist instruction

The material presented in this chapter suggests several ways in which instruction on reproductive biology and human origins can counter genetic essentialism, such as defining sex and gender and using the two terms separately; balancing the number of images of women and men and depicting them equally in various activities, environments, and poses; and representing science as a field that includes people of diverse genders and ethnicities. Other techniques include replacing the passive voice with the active voice in descriptions of female bodies and gametes; describing fertilization as the fusing or joining of cells; and providing balanced amounts of equally accurate information on female and male topics without value-laden language. Instead of the male-first convention, terms and topics should alternate between male-first and female-first but avoid switching to all female-first or all female pronouns, which would reinforce the association of females with reproduction (Campo-Engelstein and Johnson 2014).

With regard to sex determination and differences, the anti-essentialist approach summarized in Figure 4.3 involves discussing environmental, genetic, and hormonal inputs that affect phenotypes; providing examples that illustrate the lability of sex in other species to help students understand that sex is not a fixed characteristic; and explaining the multiple ways of defining sex, the homology of hormones and reproductive organs, and the continuous variation in cognitive and behavioral traits within sexes and genders along with the overlap in ranges between categories. Exploring sociocultural and historical factors behind inequalities in education, family life, and the workplace can further prevent concepts about reproductive traits and sex-linked genetic conditions from feeding into essentialist beliefs about behavioral, psychological, and cognitive ones, and carries the side benefit of expanding students' understanding of the range of scientific inquiry to include the topic of gender. Finally, teaching about evolution should address entrenched assumptions about sex roles and social organization in the deep past that lend weight to the separate

**Anti-essentialist approach to teaching about reproductive biology and human evolution**

- sex and gender distinguished and defined
- complexities of phenotypic sex
- homology of reproductive parts and hormones
- equal accuracy of information regarding female and male reproductive biology
- no male-first convention
- gender-balanced images in terms of numbers and activities presented
- images that reflect diversity of sexes, sexualities, genders
- continuous variability in cognitive and behavioral traits within groups and overlap between
- sociocultural, political, and historical factors affecting scientific projects and workforce
- explanations of evolution that contradict pervasive assumptions about sex roles in the deep past

*Figure 4.3* Anti-essentialist education concerning reproductive biology and human evolution.

characteristics attributed to women and men and which are the subject of the following two chapters.

An accurate understanding of reproductive biology, sex determination, and human evolution is needed to give students useful tools for understanding and appreciating themselves and others. An anti-essentialist approach encourages interest and sense of belonging in science on the part of women and students from underrepresented groups. It works together with anti-essentialist genetics instruction to refute misconceptions about human differences that distort perception and justify discrimination and inequality.

## Notes

1 $n = 5{,}618$.
2 $n = 3{,}280$.
3 See Cassese and Bos (2013) for an analysis of images in US political science textbooks.
4 $n = 333$ (sample drawn from 2,164 resources that met the criteria for inclusion).
5 Date range 2005–2009.
6 Date range 2010–2018 plus one from 2002.
7 The books were drawn from sample of 132 schools.
8 Date range 2009–2016.
9 For instance, with respect to first intercourse, boys were said to experience performance anxiety and be subjected to unpredictability in erection and ejaculation, whereas girls were said to experience fear that intercourse will hurt and that engaging in it will change how the boy sees them.

10 $n$ = 233, biology of sex course; $n$ = 135, introductory course for non-majors; $n$ = 171, course for biology majors. Date range of books: 2015–2019.
11 Date range n/a. The one history book is from Cyprus.
12 One exception was a picture in a Finnish book of two women working with a chimpanzee.
13 Date range n/a. Books from five other countries were excluded from the analysis: those from Burkina Faso because it was not included in the project at the time; those from Finland and Hungary because there were no such images; and those from Portugal and Tunisia because the textbook did not include a section on human origins although there was a chapter on human evolution.
14 Four polls between 1981 and 2005 found that 47–76% of respondents believed that creationism should be taught alongside evolutionary science in public schools. Five polls between 1999 and 2005 yielded a rate of 33–44% who preferred creationism over evolution (Pobiner 2016).
15 $n$ = 5,706.
16 Teaching of evolution as well as inclusion of human origins changed over the research period and varied across countries by grade level and course type (biology, psychology, history, philosophy), and, in Burkina Faso, between French-model and other schools.
17 In response to a second question about factors related to species evolution (such as natural selection, chance, environment), 8% of teachers in France ticked either the box that said God had great importance or some importance, compared to 95% or more in Lebanon, Senegal, Tunisia, Morocco, and Algeria.

## References

Aivelo T, Neffling E, Karala M. 2024. Representation for whom? *Journal of Biological Education* 58(2):297–311.
Anifandis G, Messini C, Dafopoulos K, Sotiriou S, Messinis I. 2014. Molecular and cellular mechanisms of sperm-oocyte interactions opinions relative to in vitro fertilization (IVF). *International Journal of Molecular Sciences* 15(7):12972–12997.
Bergström A, Tyler-Smith C. 2017. Paleolithic networking. *Science* 358(6363): 586–587.
Biemmi I. 2011. *Educazione sessista*. Rosenberg & Sellier.
Caldwell EF, Wilbraham SJ. 2018. Hairdressing in space. *Journal of Science & Popular Culture* 1(2):101–118.
Campo-Engelstein L, Johnson NL. 2014. Revisiting "The fertilization fairytale." *Cultural Studies of Science Education* 9(1):201–220.
Casey K, Novick K, Lourenco SF. 2021. Sixty years of gender representation in children's books. *PLoS One* 16(12):e0260566.
Cassese EC, Bos AL. 2013. A hidden curriculum? *Politics & Gender* 9(2):214–223.
Donovan BM, Stuhlsatz MA, Edelson DC, Buck Bracey ZE. 2019. Gendered genetics. *Science Education* 103(4):719–749.
Donovan BM, Syed A, Arnold SH, Lee D, Weindling M, et al. 2024. Sex and gender essentialism in textbooks. *Science* 383(6685):822–825.
Dunk RD, Malmquist SJ, Prescott KK, Ewell SN, Henning JA, et al. 2024. How do students critically evaluate outdated language that relates to gender in biology? *CBE—Life Sciences Education* 23(2):ar24.
El Takach S, Yacoubian HA, Hagop A. 2020. Science teachers' and their students' perceptions of science and scientists. *International Journal of Education in Mathematics, Science and Technology* 8(1):65–75.

Elgar AG. 2004. Science textbooks for lower secondary schools in Brunei. *International Journal of Science Education* 26(7):875–894.
Fuselier L, Eason PK, Jackson JK, Spaulding S. 2018. Images of objective knowledge construction in sexual selection chapters of evolution textbooks. *Science & Education* 27(5–6):479–499.
Georgadaki K, Khoury N, Spandidos DA, Zoumpourlis V. 2016. The molecular basis of fertilization. *International Journal of Molecular Medicine* 38(4):979–986.
Giacomini M, Rozee-Koker P, Pepitone-Arreola-Rockwell F. 1986. Gender bias in human anatomy textbook illustrations. *Psychology of Women Quarterly* 10:413–420.
Good JJ, Woodzicka JA, Wingfield LC. 2010. The effects of gender stereotypic and counter-stereotypic textbook images on science performance. *The Journal of Social Psychology* 150(2):132–147.
Graves Jr JL. 2015. Great is their sin. *The Annals of the American Academy of Political and Social Science* 661(1):24–50.
Hannover B, Kessels U. 2004. Self-to-prototype matching as a strategy for making academic choices. *Learning and Instruction* 14(1):51–67.
Kerkhoven AH, Russo P, Land-Zandstra AM, Saxena A, Rodenburg FJ. 2016. Gender stereotypes in science education resources. *PLoS One* 16(11):e0165037.
Lee JFK, Collins P. 2010. Construction of gender. *Journal of Gender Studies* 19(2):121–137.
Loughlin AT. 2018. Penetrate science. *Working Papers of the Linguistics Circle* 28(1):60–77.
Martin E. 1991. The egg and the sperm. *Signs: Journal of Women in Culture and Society* 16(3):485–501.
McCabe J, Fairchild E, Grauerholz L, Pescosolido BA, Tope D. 2011. Gender in twentieth-century children's books. *Gender & Society* 25(2):197–226.
Metoyer AB, Rust R. 2011. The egg, sperm, and beyond. *Women's Studies* 40(2):177–205.
Moraïs S, Winkler S, Zorea A, Levin L, Nagies FSP, et al. 2024. Cryptic diversity of cellulose-degrading gut bacteria in industrialized humans. *Science* 383(6688):1197.
Moser F, Hannover B. 2014. How gender fair are German schoolbooks in the twenty-first century? *European Journal of Psychology of Education* 29(3):387–407.
Nandi A, Halder T, Das T. 2024. Gender stereotypes and bias in school mathematics textbooks. *Current Perspectives in Educational Research* 7(2):36–50.
Parker R, Larkin T, Cockburn J. 2017. A visual analysis of gender bias in contemporary anatomy textbooks. *Social Science & Medicine* 180:106–113.
Plutzer E, Branch G, Reid A. 2020. Teaching evolution in US public schools. *Evolution: Education and Outreach* 13(14):1–15. https://doi.org/10.1186/s12052-020-00126-8
Pobiner B. 2016. Accepting, understanding, teaching, and learning (human) evolution. *American Journal of Physical Anthropology* 159(S61):232–274.
Quessada MP, Clément P. 2011. The origins of humankind. In Yarden A, Carvalho GS, editors. *Authenticity in biology education. Benefits and Challenges*. ERIDOB & CIEC. pp. 295–307.
Quessada MP, Clément P, Oerke B, Valente A. 2008. Human evolution in science textbooks from twelve different countries. *Science Education International* 19(2):147–162.
Sadava D, Hillis DM, Heller HC, Hacker S. 2021. *La nuova biologia.blu PLUS*. 2nd ed. Zanichelli.

Sovič A, Hus V. 2015. Gender stereotype analysis of the textbooks for young learners. *Procedia-Social and Behavioral Sciences* 186:495–501.
Stuhlsatz MA, Buck Bracey ZE, Donovan BM. 2020. Investigating conflation of sex and gender language in student writing about genetics. *Science & Education* 29(6):1567–1594.
Whitaker ED, Baccolini R. 2024. Heroes and helpmeets. *Science & Education* 33(1):271–296.
Wiber MG. 2006. *Erect men/undulating women*. Wilfrid Laurier University Press.
Wood S, Henning JA, Chen L, McKibben T, Smith ML, et al. 2020. A scientist like me. *Proceedings of the Royal Society B* 287(1929):20200877.
Yacoubian HA, Al-Khatib L, Mardirossian T. 2017. Analysis of the image of scientists portrayed in the Lebanese national science textbooks. *Science & Education* 26(5):513–528.

# 5 Sexes, genders, and sources of difference

The last chapter examined how biology textbooks reinforce common misperceptions about sex and gender. Together with the use of these words as interchangeable terms, simplified explanations of sex determination and physical differences imply that sex causes gender-divergent cognitive and behavioral traits as well. Discussions of human evolution offer an ultimate rationale for sex/gender essentialism through assertions about purported sex differences in productive and reproductive roles in the deep past.

This chapter offers an anti-essentialist alternative by examining sex and reproduction in humans and other species, physical sex differences and similarities, and brain plasticity. It reviews studies on brain function and behavior which confirm that the brain and mind do not come in separate female and male versions. Analyses of test scores and academic performance, the effects of hormones on cognition, and emotional expression likewise refute essentialist beliefs about gender and cognition. Finally, this chapter reviews the variability in sexes, sexualities, and genders within and between societies, as well as diversity in gender concepts and roles. This information complements the analysis in Chapter 6 of gender fallacies embedded in narratives about human evolution.

## Sex as a master category

While every culture recognizes that babies are born from females and not males, a dichotomous vision of sex overlaid by a gender binary is not the only way to catalog people or structure social roles and relationships. It may seem natural to much of the world today, given the status of sex as the master category (with race close behind). Sex is a required piece of information on birth certificates, driver's licenses, and airplane tickets. Women who work as physicians are called "lady doctors," while men who work as nurses are "male nurses." The adjectives are left off where people's profession aligns with cultural expectations: men who fly planes are just "pilots"; women working in primary schools are simply "teachers."

DOI: 10.4324/9781003639923-5

The habit of categorizing people first and foremost by gender as a proxy for sex causes similarities within groups and differences between them to be automatically attributed to sex. For instance, the presumed sex difference causing women to have more acutely sensitive fingertips turns out to be a question of fingertip size: men with smaller hands have greater fingertip sensitivity than women with larger hands (Peters et al. 2009). Similarly, the idea that women are chattier than men does not match the evidence: researchers in the US and Mexico found that both women and men averaged around 16,000 words per day (Mehl et al. 2007). A larger study found a small difference in favor of women, due mainly to more words per day among those in the early and middle adulthood range, although older women's average was lower than men's while very wide individual differences limit the generalizability of the findings (Tidwell et al. 2025).[1]

The color-coded environments created for girls and boys reflect the essentialist idea, rooted in an evolutionary storyline, that girls like pink because ancestral females needed to be able to distinguish berries against a green background whereas boys like blue because males needed to read the sky for clues to its propitiousness for hunting. Children's preferences contradict these assumptions. Both boys and girls prefer red through the first two years of life – the period in which they need to locate their mother's nipple – after which their preferences diverge in line with cultural expectations (Jadva et al. 2010). In societies outside the influence of global culture, children do not exhibit a sex difference in preference for pink, as shown by a comparison of 4–11-year-olds in an Australian city versus three small villages in Peru, Vanuatu, and Republic of Congo (Davis et al. 2021).

Assumptions about the characteristics of human females and males find their way into animal studies, shaping how behavior is perceived and interpreted. For instance, until researchers tracked koalas with GPS collars, it was assumed that males had larger ranges than females and made nighttime calls directed at other males. Instead, there was no difference in the size of female and male ranges, and male nocturnal vocalizations served more to attract females from their treetop nests during breeding season than communicate territorial information to other males (Ellis et al. 2011).

The symbolic opposition between male and female aligns with an interconnected array of contrasting values and concepts in Euro-American culture: strong/weak, hard/soft, active/passive, subject/object, mind/body, reason/emotion (or thought/feeling), clean/unclean, right/wrong, true/false, civilization/nature. Further oppositions are related to biophysical properties and entities, lending the whole system the air of naturalness: up/down (or top/bottom), right/left, sun/moon. Each of the first terms has superior connotations to the second. Even the sciences are described as "hard" and "soft": physical and natural sciences aligned with male; social sciences coded as female (see Shapin 2022). The "hard" sciences themselves fall along a declining prestige range from abstract, "clean" physics to earthy, messy biology. Within biology, genes are associated with male, agency, and truth, whereas environments are aligned with female, context, and uncertainty.

These associations are built into widespread, tenacious assumptions about the characteristics of women and men. They are not dictated by human biology, nor do they manifest themselves in universal gender expectations or traits, as seen below.

## The diversity of sex

The last chapter reviewed how biology textbooks portray sex determination as a straightforward process whereby gametes carrying either an X or Y chromosome join to create a new cell with two Xs or an X and a Y. Presenting sex as binary conforms to the traditional belief that sex determines both opposed gender traits and heteronormative sexuality. These ideas clash with the realities students encounter in everyday life, in which they observe that sexuality is variable and very few people consistently behave in line with gender expectations. Since intersex conditions are less common, many students may not be aware that biological sex also comes in more than two forms. An expanded and more accurate explanation of sex in biology textbooks would help fill in the blanks and render the material not only more interesting but also more believable.

To begin with, modes of reproduction in the living world are highly variable and context-dependent (see Fuentes 2025). Asexual reproduction is comparatively safe and efficient for short-lived species that can respond to environmental change through mutation, whereas longer-lived species trade the effort and risk of sexual reproduction for the genetic diversity that comes from shuffling genes during the production and meeting of gametes. Among sexually reproducing species, it is not always the males that produce gametes with two different sex chromosomes: for instance, female birds have ZW chromosomes, males ZZ, such that the sex of offspring depends on whether they receive a Z or W from the mother. While the human Y chromosome is a third the size of the X (with only 27 protein-coding genes), in the white campion plant (*Silene latifolia*) and related species it is larger than the X and contains hundreds of genes. The number of chromosomes varies as well. In platypuses there are five types of X and five types of Y: males have five pairs of the different Xs and Ys; females five pairs of Xs. In *Caenorhabditis elegans* roundworms, XO individuals produce male gametes only, whereas XX worms produce both sperms and eggs and can self-fertilize. A single gene determines the sex of asparagus plants: homozygous recessive individuals are females, whereas heterozygotes are males.

Instead of two sexes, many species of fungi have thousands of mating types. While species of plants, algae, and microorganisms that reproduce asexually have no sexes, other species alternate between asexual and sexual reproduction. These include aphids, some bees and wasps, whiptail and other lizards, nematode worms, and freshwater snails. The alternation – like that between sexes which occurs in many sexually reproducing species of insects, reptiles, plants, and fish – depends on environmental cues such as temperature, ocean currents, parasites, gases in soil, and seasonal light levels. Species that can produce either type of gamete include most snail and slug species, and many of the approximately 2,000 species of ray-finned cichlid fish, including clownfish, sea bass, and wrasses. Instead of switching back and forth, sex in American alligators is set permanently by the temperature of the mound of decaying plant matter in which eggs are laid (or the part of the mound where a specific egg is situated). There is a critical period starting about half-way to hatching: cool results in females, hot in males, and middle temperatures (30–34 degrees Celsius) in a mix of the two. For many lizard and turtle species it is the opposite: males develop at cool temperatures, females at hot, and a mixture in the middle range. Many flowering plants not only have both male and female reproductive organs but also self-fertilize. Alternatively, reproduction without fertilization through parthenogenesis occurs frequently in insects and has been documented in some species of reptiles, birds, and sharks.

Parenting behavior likewise varies and can change over the lifetime depending on circumstances. For instance, while albatross couples normally consist of a female and male that care for and hatch the egg together, on some Hawai'ian islands males are scarce and up to a third of the couples are long-term female-female pairs that take turns laying an egg. The proportion of their eggs that hatch is lower than for female-male pairs, but they are equally successful at bringing chicks to fledging age (Young et al. 2008).

In sum, living things reproduce in a great many ways. Human sexual/reproductive development likewise fails to follow a uniform pattern and involves more chromosomal combinations than the XX/XY model suggests. In human embryos, genes that will later specialize by sex are expressed equally in females and males, whose cells can develop along either path. Beginning in the earliest days of development, precursor cells to female and male gonads migrate to the genital ridge. The sex-determining region (SRY) of the Y chromosome initiates a series of events that causes differentiation along the male pathway: female XX mice which have had the region inserted into their genome develop into males. However, male development depends also on genes on the X chromosome: altering the expression of genes on the X chromosome causes mice with no Y chromosome to develop testes and produce sperm, although they do not develop full reproductive competence (Yamauchi et al. 2016).

The SRY stimulates production of testis-determining factor, which activates the *SOX9* gene on chromosome 17, whose Sox9 protein is a transcription factor that activates fibroblast growth factor 9 (Fgf9). The latter induces cells of the genital ridge to differentiate into Sertoli cells rather than the granulosa cells of female development, while other genes support the further formation of the testes. In addition, the SRY blocks the anti-testis activity of the DAX-1 protein coded by the X-chromosome *NR0B1* gene; without DAX-1, or Fgf9, an XY fetus develops into a female.

At around two months, the genital ridge differentiates into either testes or ovaries, with associated structures (seminal vesicles, epididymis, vas deferens, penis; or fallopian tubes, uterus, vagina). If any of the male-prompting elements is missing, an XY embryo develops along the female path. However, female development is not automatic and requires multiple inputs and both Xs: the 10% of XO females who survive embryonic life remain infertile. Reproductive competence in males also requires genes on the X chromosome, which, together with genes on the Y, are necessary for sperm production by spermatogonia in the testes. After embryonic development, it is genes on the X and autosomal chromosomes that regulate the hormonal processes involved in secondary sex characteristics.

Not only do reproductive organs originate in homologous embryonic structures, but estrogens and androgens are chemically related hormones built from cholesterol and other precursor molecules, with the four-carbon ring structure of steroid hormones. These hormones circulate in bodies of all sexes. Humans share them with reptiles, birds, mammals, and, with slight variations, fish and insects. They are synthesized not just by the ovary and testis but diverse sources, including the adrenal cortex, arterial walls, liver, brain, placenta, breast, adipocytes, and skin. So-called male and female hormones play important roles in all bodies. For instance, estrogen is needed to halt the growth of the long bones after puberty. Insufficient amounts leave girls and boys with elongated limbs. Lack of estrogen or its receptors prevents the production of both eggs and sperms and causes multiple sexual and reproductive defects. In male fetuses, both testosterone and estrogen are required for normal penis development.

There are many other illustrations of shared reproductive anatomy and physiology between males and females. For instance, the hypothalamus and pituitary glands of a castrated male laboratory animal will switch from a continuous to cyclical pattern of hormone secretion under the influence of administered estrogen. In humans, the female-associated hormones, oxytocin and prolactin, facilitate trust and interpersonal relationships in all individuals. Lactation is not limited to females, for it can occur in males in response to nipple stimulation, starvation, and medical treatment (see Geller 2009).

Where the question is reproductive competence and its development beginning with gametes carrying X or Y chromosomes, a binary vision of sex is appropriate and accommodates the shared biology described above. However, a sex spectrum is more accurate given that many people fall outside the binary, whether because of chromosomal anomalies, hormonal variations, or developmental factors. Reproductive structures may remain underdeveloped; there may be a male or female phenotype together with nonfunctional reproductive organs; or there may be a mixture of features or features that do not align with chromosomal sex. The proportions at birth vary across societies and remain disputed and uncertain due to uneven criteria and quality of reporting, while those which regard adults are distorted by the fact that individuals born with ambiguous characteristics are often treated medically and surgically to fit into the female or male category (see Hegarty and Smith 2023).

While the frequency range leaves room for opposition to the concept of a sex spectrum (see Bogardus 2025; Elliott 2023), the latter best describes a distribution that is strongly but not exclusively bimodal. At the high end, Fausto-Sterling (2000) reports that 1.7% of births have chromosomal or other variations that result in nonstandard anatomical features and/or hormone levels, which Dunk et al. (2024) point out is equivalent to the worldwide frequency of red hair (1–2%). Lower estimates concern specific conditions or are based on stricter criteria, such as Sax's (2002) report of the frequency of intersex at 0.018%.

Females with a missing X chromosome (XO) or cells with one X and a partial second X (Turner Syndrome) develop diverse symptoms of varying severity, including heart conditions, short stature, and, as noted above, infertility. Conditions in which there is an extra chromosome include Triple X Syndrome, XXY or Klinefelter Syndrome, and XYY or Jacobs/Superman Syndrome, and result from a lack of separation of the sex chromosomes during the first phase of meiosis or chromatids during the second. During crossing-over, a section of the Y chromosome may be transposed onto the X, causing an XX female to develop male characteristics (XX Male Syndrome); in XY males, its displacement causes the development of female characteristics.

Besides chromosomal anomalies, hormonal and other disorders can interfere with sexual development or cause masculinization or feminization opposite to chromosomal sex. Congenital adrenal hypoplasia, caused by a mutation or deletion of the *NR0B1* gene mentioned above, results in underdeveloped reproductive structures in XY males. Contrariwise, XX females with congenital adrenal hyperplasia develop male characteristics. Five-alpha-reductase deficiency syndrome causes XY males to develop ambiguous genitalia; female genitalia develop in XY males with androgen insensitivity syndromes such as Swyer Syndrome. Mosaic characteristics can also arise from chromosomal anomalies due to fertilization by two sperms, loss of a sex chromosome by a group of cells during development, or exposure to exogenous hormones and chemicals that mimic testosterone.

The variability described above shows that the binary model cannot adequately describe the diversity of human sexes. Moreover, male and female reproductive structures originate from the same embryonic source, while developmental and physiological processes work in generally the same way regardless of sex. At the same time, there are meaningful differences between females and males, as discussed below.

## Anatomical and physiological differences

Sex differences fall into two types: dichotomous and continuous. Dichotomous traits stem from sexual and reproductive structures and functions without overlap between females and males, yet even these traits vary within categories due to the gene-environment-experience processes discussed in Chapter 3. For instance, the age at which girls begin to menstruate varies across historical periods and between and within societies (see Eaton et al. 1994). Women in foraging societies reach menarche at 17–18 years, rather than 12–13 as in much of the world today. European women underwent a shift between these two age levels in step with industrialization, as diet, physical activity, and infectious disease patterns changed. Elite women had long reached menarche earlier, thanks to the same factors. The age at menopause meanwhile increased from around 40–50 or more years. Today, chemicals that mimic hormones and which are ubiquitous in the environment are lowering the average age at menarche even further.

For another example, despite the widespread assumption that the decline in estrogen and termination of childbearing with menopause faithfully bring hot flushes, depression, and bone weakness, symptom reporting varies widely across and within societies. These symptoms came to be associated with menopause beginning a century ago, as Western medicine transformed the climacteric from a transitional phase for all people (and one that protected older women from pregnancy) into a pathological state for women only. In practice, most women even in Europe and North America go through menopause without serious disturbances, and they are less likely to experience depression than younger women. Elsewhere, symptoms vary in magnitude and type due to multifactorial influences that illustrate the concept of "local biologies" (Lock and Kaufert 2001). For instance, Mayan women experience the same reduction in estrogen levels as North American women and lose bone mineralization to an even greater degree, yet they do not report night sweats or hot flushes and rarely experience bone fractures (Beyene and Martin 2001). Estrogen levels vary across women due to hormonal contraceptive use, childbearing and breastfeeding history, smoking, exercise, and diet. For Asian women, the last of these includes phytoestrogens in soybeans and teas that may modulate temperature-regulation symptoms, as may a lifestyle that includes daily physical activity, avoidance of smoking, and rare consumption of alcohol or caffeine. Japanese women's main complaint is shoulder stiffness, stemming from their physically and psychologically demanding duty to care for older relatives (Lock and Kaufert 2001).

The concept of local biologies points to the need for caution in generalizing about all humankind from samples of humans from a particular society and/or point in time. Studies based on Europeans and Americans dominate the scientific literature, even though their way of life is unlike many others. Impacts are highlighted further below and in the next chapter.

While the X-linked genetic variants that cause red/green colorblindness and diseases such as hemophilia appear discontinuous because they mainly affect males (see Chapter 4), there is a range of symptoms among females as well. Female heterozygous carriers may be asymptomatic or moderately affected depending on which X is inactivated in which cells and which cell type. Symptoms are more severe among the rare homozygous females.

Among continuously variable traits, body size is among the most apparent. The difference in height between human females and males ranges from 88% to 96% (Wolfe and Gray 1982), with the greatest sexual dimorphism in size among the taller populations of temperate zones and the least among the shorter populations of equatorial and arctic zones.[2] Compared to other primates, the degree of dimorphism in humans indicates an evolutionary history of mild male-male competition (see below, and Jolly 2001). Closely related species with more intense male-male competition are more divergent in size: gorilla females are half the size of males, while chimpanzee females are three-quarters the weight of males although equally tall among common chimps and 5–10% shorter in bonobos. Implications of size, strength, and endurance for sex roles in prehistory are discussed in Chapter 6.

Larger size is not the norm across mammal species, as shown by analysis of a sample of 400 species (5% of the total) that yielded the following percentages: same-size females and males in 39% of species; larger females in 16%; larger males in 45% (Tombak et al. 2024). Larger males were most common among carnivores, larger females among bats (accounting for one-half of species, with same-size sexes in one-third and larger males in one-sixth). Aside from bats, which carry infants while flying, female mammals are more constrained in growth than males due to the energy requirements of pregnancy and lactation. A few primate species have larger females, including potto monkeys and some langur and gibbon species. Those with same-sized males and females include small-bodied species such as tamarins and marmosets, which form long-term bonds and participate equally in parenting. Among large-bodied species such as gorillas and chimpanzees the males are considerably larger, compete more for mating opportunities, and provide less caregiving to the young.

Across mammal species, males are subject to higher mortality rates than females at all age levels. In humans, for most of recorded history and in the most impoverished countries until only the last few decades, gender inequality overwhelmed this female biological advantage for much of women's life cycle. This is because female mortality rates exceed male rates from childhood through the reproductive years wherever life expectancy rates are below 50 years. Today, while female mortality rates are lower than male through most of the life course, they converge at the oldest ages. There are nearly equal numbers of women and men in the world's "longevity blue zones" filled with ultra-centenarians, such as Okinawa, Japan, and the mountainous interior of Sardegna, Italy (see Pes et al. 2015).[3]

The female mortality advantage is linked to differences in immune responses stemming from two rather than one X chromosome. Together with variable X inactivation, the 25–30% of the approximately 800 immunity-related genes on the X that escape inactivation bring females reduced susceptibility to respiratory syndromes, infectious diseases, and infection-related cancers, with lower severity and lower mortality rates (see Sharma et al. 2025). Contrariwise, the enhanced immune activity can turn upon self-cells when triggered by genetic, environmental, and hormonal processes, rendering females far more susceptible to autoimmune diseases such as lupus and scleroderma, along with long-term sequelae of infection which are themselves a form of autoimmune response. Males have the benefit of higher levels of testosterone, which protects against inflammation through the hormone's immunosuppressive effects.

Many kinds of sex difference arise from divergences in hormone levels. Together with genetic mechanisms, hormones affect the production of subtypes of immune cells and their presence and activity in diverse tissues and organs. Hormones play a role in the migration and interaction of immune cells by way of signals to nonimmune cells, such as skin and fat cells. Their signaling functions with respect to immune cells distributed throughout the body, including the nervous system, contribute to sex differences in the experience of pain that, however, are not categorical but rather indicate male-typical and female-typical pathways (see Sorge and Strath 2018). The spectrum of sexes, the overlap in hormones between people of all sexes, and the change in hormone levels over the lifetime lead to combinations and alternations in individual people. In old age, sex differences in pain mechanisms fade away. For all age groups, psychological factors related to gender constructs and expectations confuse the picture through their effects on pain perception.

In addition to greater vulnerability to infectious disease, and to more severe illness, males are less protected against damage and injury to tissues and organs including the brain due to lower levels of progesterone. Males have higher risks of hypertension and cardiometabolic disorders due in part to Y-chromosome genes that influence adaptive immunity, including T-cell activity (Khan et al. 2019), as well as differences in fat distribution and metabolic functions (see Chapter 6). Males are several times more likely than females to develop neurodevelopmental disorders (NDD), including attention-deficit and autism-spectrum disorders. Reasons for the difference in rates of NDD include X-linked conditions, Y chromosome genes, and the effects of fetal testosterone on the developing brain. The rapid rise in the prevalence of these disorders over the past few decades suggests that environmental factors also play a role. In pregnant mice, exposure to air pollution together with stress (induced by restricted materials for building a nest) results in the same effects on male but not female fetuses as infection-induced maternal immune activation: dysregulated macrophage-like microglia in the brain that fail to properly pare down the superabundance of synaptic connections present at birth, especially in the anterior cingulate cortex that is central to processing social cues (Block et al. 2022). These same dysregulated cells are characteristic of human patients with NDD. Moreover, experimental inhibition of microglial phagocytosis in the mouse anterior cingulate cortex results in the same aberrant communication and social behavior in early life and over the long term as the pollution-plus-stress combination during fetal life.

The sex differences described above and summarized in Figure 5.1 are important and have significant implications for health and wellbeing, yet they are not absolute: autoimmune diseases, NDD, and pain affect people of all sexes. Body size falls along overlapping ranges. As shown below,

Types of variation in physical sex differences

| dichotomous | continuous |
|---|---|
| • sexual and reproductive anatomy and functions (e.g. menstruation) | • stature<br>• size<br>• mortality rates<br>• sex-linked genetic traits (e.g. red/green colorblindness, hemophilia)<br>• gene-immune-hormone processes (e.g. disease susceptibility and severity, pain perception, neurodevelopmental disorders) |

*Figure 5.1* Dichotomous versus continuously variable physical sex differences.

these physical differences dictate neither sexuality nor gender identity and are not the source of gender differences in cognitive and behavioral traits.

## Brain size, morphology, functioning

The popular belief that men have larger brains than women and are therefore more intelligent defies both logic and evidence. It requires affirming that tall women are smarter than short men given that brain size rises roughly in proportion to height. As noted in Chapter 1, brain size does not correlate with intelligence. Moreover, uncertainty remains over whether male brains truly are bulkier after correcting for body size, due to confounding factors such as how accurately the brains in a sample have been classified by sex and the effects on brain volume of age, disease, and plasticity in response to use.

Thanks to the failure of the volume argument for greater male intelligence, the search is on for sex differences in features such as neuronal connectivity or the size of specific brain regions. Where differences are found, caution is needed since a sample of individuals in one population cannot be assumed to represent all females and males in the same society, much less all humankind. Most fatally for the presumption of separate male and female brains, nearly all brains combine features more typical of males with ones more typical of females. This has been shown by an analysis of multiple datasets on more than 1,400 brains, including MRI scans that measured the volume of specific brain regions (gray matter, or densely packed nerve cell bodies) and nerve cell fibers that transmit signals throughout the brain (white matter), as well as diffusion tensor imaging of the distribution of nerve cell fibers indicating connectivity between regions (Joel et al. 2015). For all the features in the categories of gray matter, white matter, and connectivity, there was considerable overlap in the distribution of female and male values. Those with the least amount of overlap were singled out, the data points arranged along a scale from male-typical to intermediate to female-typical with each category representing one-third of the range.

Analysis of subsets of brains demonstrated that nearly all individual brains were mosaics of features more common in females, features more common in males, and features intermediate between the two.[4] There were far more brains with substantial variability in features than internal consistency. Across multiple samples, from 0% to 5.9% of brains had all their features fall into one category (female, intermediate, or male). One-fifth to more than one-half of brains showed substantial variability in male-typical and female-typical features. Females and males were equally likely to show non-consistent features.

The study shows that brains do not exist in separate female and male forms and that traits more typical of one sex or the other vary independently across features. The researchers also confirm that these findings match people's gender-inconsistent behavior, interests, and attitudes. Based on datasets on thousands of individuals, they report that for the seven traits with the greatest gender difference, there was substantial variability in 59% and 70% of individuals.[5] All the internal consistency occurred in the intermediate range of values, and accounted for 1.8% and 0.1% of individuals, respectively. Finally, an analysis of 10 highly gender-stereotyped behaviors of college students, such as playing video games (with *d* scores falling between 1.00 and 2.02, discussed below), yielded similar results.[6] There was substantial variability in 55% of individuals and internal consistency in 1.2%. Taken together, these analyses clearly demonstrate that neither brains nor cognitive traits segregate by sex.

Even the variability in the brain features identified in the above study cannot be taken as a reflection of inherent sex differences, for the way the brain is used affects its morphology and functioning, as does injury or disease. For instance, damage to the parts of the brain that process speech and vision can cause these tasks to be shifted to other areas. Starting in early life, synaptic connections are pared down in response to experience, as noted above. Throughout life, task-oriented training affects vascularization, neuronal branching and interconnections, and the production of neurotransmitters.

Experiments with birds and mammals show that extensive activity involving orientation in space affects the size of the hippocampus, which deals with visuospatial relationships (see Yaskin 2011). Seasonal restrictions or experimental manipulation of the environment, range size, or ability to store and retrieve food results in a smaller hippocampus, whereas restoration of normal conditions brings an increase in volume. Among humans, taxi drivers have larger-than-average hippocampi, in proportion to years of driving experience (Maguire et al. 2000). Similarly, several brain areas respond to daily Zen meditation. Parts of the cerebral cortex grow thicker including the insula, which is involved in motor control, self-awareness, and various cognitive tasks. There is increased development of the nearby anterior cingulate cortex that, as noted above, processes cognitive and emotional functions, including the subjective experience of pain. As with taxi drivers, the scale of the response is proportional to the dose of behavior: more years of training bring increased brain tissue thickening and greater pain tolerance (Grant et al. 2010).

The fact that humans are born with a brain one-fourth its adult size (300–400 versus 1,300–1,400 grams) indicates that there is ample scope for extra-uterine influences on development. Children grow up in a context of intensive social interaction, imitation, and learning. Gender beliefs nudge their activities and interests along culturally paved paths. If a high proportion of females and males follow these paths, the training can yield average differences in the size, shape, and electric activity of various parts of the brain, creating the illusion of inborn anatomical disparities.

The process of setting girls and boys on different trajectories begins even before they are born through parental choices regarding the baby's room colors and decorations, clothing, and equipment. From the moment they are born, babies are perceived and treated in line with the way they are presented, as shown by experiments in which they are dressed opposite to their sex. Girls are seen as delicate and pretty, boys as strong and active, leading presumed girls to be treated gently while boys are handled roughly. When babies are given a physical challenge such as climbing up an inclined plane, their parents mistakenly expect girls to be less capable. Despite the absence of a gender difference in toy preferences among infants, separate types of toys are offered to girls and boys (Eliot 2010).

Male babies are born slightly larger than female babies, thanks to the growth-spurring effects of testosterone in utero. Bodily growth takes away from neurological development, leaving boy babies with less acute vision and hearing, and less developed fine motor, communication, and memory skills. Boy babies tend to be more irritable and consequently fussier, and they have a harder time soothing themselves. This can lead parents to pay more attention to boys, including giving them more physical contact, visual stimulation, and emotion management. Tendentially, more placid females are more likely to be left alone and to be engaged verbally instead of physically. Although the developmental differences fade over the first year, parents may continue to emphasize physical activity in boys more than girls, with enduring impacts on their physique, athletic ability, interests, and brain (Eliot 2010).

According to a common misconception, the menstrual cycle affects the female brain and therefore women's cognitive abilities. To test the evolutionistic assumption that estrogen reduces visuospatial ability but enhances multitasking skills in line with the alleged demands of ancestral hunting versus gathering/mothering, Leeners et al. (2017) followed 18–40-year-old women over two full menstrual cycles. The women were healthy, not pregnant, and had not lactated in the previous six months. Blood samples and three cognitive tests were taken twice at each of four phases of the menstrual cycle: menstrual, pre-ovulatory, mid-luteal, and premenstrual.[7] The tests were administered on a touch screen computer: one each for visuospatial working memory, divided attention, and cognitive bias/control.[8] Performance on none of the three tests was altered by the menstrual cycle, although there were some differences in individual women in the first cycle that were not repeated in the second.

The idea that males are especially endowed with visuospatial abilities for hunting is connected to the theory that they are more visually driven than females due to upright posture. Upright posture allegedly shifted the male gaze from the sexual swellings that signal estrus in nonhuman primates to visual breast tissue (attributed to sexual selection) and the waist-to-hip ratio as gauges of youth and fecundity. This scenario conflicts with the fact that female and male humans have the same visual equipment and respond to visual stimuli the same way (Geller 2009). Visual breast tissue ranges from minimal to substantial and expands with lactation, when conception is unlikely. Waist-to-hip ratios vary considerably across and within societies, as well as in relation to physical fitness. In addition, the sexualization of the female breast is far from a cultural universal (Dettwyler 1995), and, where the breast is sexualized, larger breasts are not necessarily the most desired. For instance, until recent decades, small breasts were idealized in Brazil, and cosmetic surgery served to reduce rather than augment breast size (see Edmonds 2013).[9] Likewise, the prioritization of sight is not a cultural universal (see below), while sexual attraction depends on many senses, including smell and hearing.

The other side of the story about male interest in youth is that females are driven to prefer older males with greater control over resources. However, when asked about desired traits in a partner women and men throughout the world both rank personality, behavior, and intellectual characteristics well above either attractiveness or wealth. The personality traits women and men desire in a partner are the same ones they desire in a best friend of either the same or another gender (Laner and Russell 1998). The far lower-ranking characteristics of financial resources (and consequently older age) and attractiveness (and consequently youth) do differ in importance between women and men. Women express more interest in the former, men the latter (Walter et al. 2020), yet the magnitude of the difference varies with the strength of traditional gender beliefs (Eastwick et al. 2006) which depends, in turn, on the degree of gender inequality. Comparing across and within countries, as inequality declines, the age difference between spouses constricts, with couples roughly the same age in the most egalitarian societies (Walter et al. 2020).

Ancestral human females are even more likely to have chosen mates based on relationship and parenting skills than women in historical and contemporary societies. Wealth was not an issue since everyone had access to the necessities of life, while specialization of tasks was not possible until group size increased sufficiently during the Late Pleistocene (see Chapter 6). If the wealth-versus-youth theory were correct, then, among contemporary foraging societies, especially capable hunters should have more children than other men. They do not, although such men are highly esteemed and may attract more people to the group (Gurven and Hill 2009).

Advertising their parenting abilities would have benefited males over the course of human evolution, as indicated by the higher number of offspring of male gorillas that spend more time with infants, including ones they have not sired (Rosenbaum et al. 2018). Male gorillas adopt motherless infants and raise them with the help of other males and females, including juveniles. Their intervention prevents negative impacts on the orphans' survival, social rank, and reproductive capacity (Morrison et al. 2021).

In sum, the strands of evidence gathered above indicate that female and male brains, behavior, and cognitive traits are similar, with continuous variation and overlapping ranges. The discussion below examines how social and cultural forces rather than permanent, inborn qualities divide people's capabilities and aptitudes along gendered lines that both reflect and reinforce inequality.

## Behavior, cognition, academic performance

As noted in Chapter 3, there is no basis on which to claim that natural selection would have caused human races to develop different brains and forms of intelligence. Similarly, sexual selection is not a likely cause of separate cognitive abilities in females and males. Traits subject to sexual selection increase the bearer's reproductive success by helping it compete with individuals of the same sex (size, horns, etc.) or attract individuals of the opposite sex (plumage, complex vocalizations, etc.). These traits necessarily benefit one sex only, and they tend to be more common in males due to the higher variability in their chances of reproduction. To illustrate, peacocks carry unwieldy tails that signal good health to peahens and thereby increase the likelihood of mating; the same long tail would only burden females. No conceivable cognitive trait that arose over the course of human evolution would have benefited only males or females in reproductive competition given the lack of a sex difference in the way human ancestors lived. The recent shift to specialization (see Chapter 6), with greater male involvement in hunting and more female responsibility for gathering, would not have favored separate cognitive abilities since the activities require the same, not dissimilar skills and are carried out by both females and males. For instance, studies on Baffin Island have shown not only that indigenous women participated in hunting but also had equal visuospatial perception skills and used the same navigational techniques on land and at sea as men (Berry 1966).

Although an improbable cause of cognitive divergence, sexual selection accounts for some physical differences between human females and males, such as sexual dimorphism in size, mentioned above. The size difference, in turn, is linked to men's far higher propensity for physical aggression. Men commit 10 times as many homicides as women, along with much more interpersonal violence, warfare, and domestic violence. The age of men who commit the most violence runs from late adolescence to around 30 across societies, yet the magnitude varies enormously: for instance, the rate of homicide remains close to zero across all age groups in peaceful societies that discourage violence but spikes high in those which permit or encourage it. Comparing across and within countries, violence perpetrated by males varies with personal values intersecting with cultural beliefs about the acceptability of violence, as well as the level of socioeconomic and gender inequality: where women are more economically independent, all forms of physical aggression from partner violence to warfare are less common (see Fry 1992, 2006; Heise and Kotsadam 2015; Seddig and Davidov 2018).

Neurobiological factors that render males more likely to behave violently include the effects of testosterone on the male brain during gestation (Björkqvist 2018; Fry 2006; Staniloiu and Markowitsch 2012). These effects are evident in more male-typical physical play among girls with congenital adrenal hyperplasia, or who were exposed to androgen-mimicking drugs in utero. In the fetal brain, testosterone enhances the development of neurons in the hypothalamus, which, as part of the limbic system, is involved in processing emotion, sociality, attachment, and motivation and is connected to centers that activate muscle movement. There is also less development of brain areas involved in the mirror-neuron system that promotes empathetic and cooperative learning and behavior. Sex differences in levels of neurotransmitters, receptors for them, and the enzymes that break them down also play a role in aggression and impulse control. An example of the latter was mentioned in Chapter 3: increased aggressive and violent behavior on the part of men with low monoamine oxidase enzyme activity combined with adverse life experiences and/or events.

The greater frequency of aggressive behavior in boys than girls, both of whom have low hormone levels from infancy to puberty, points to the crucial impact of prenatal exposure to testosterone on the brain. Moreover, testosterone levels rise in people of all sexes after stressful and aggressive encounters (including tests, sports, and interpersonal conflicts), although in women baseline levels are 7–10 times lower to begin with. The role of circulating testosterone is evident in the pattern by which male violence peaks in the same age range as testosterone levels and declines after age 30 as testosterone levels fall. Comparing across males, higher baseline levels of the hormone are associated with more risk-taking, aggressive, and addictive behavior. Intimate partnership, the birth of children, and caregiving all depress testosterone secretion and increase the release of hormones that facilitate bonding and other-orientation (see Chapter 6). One of these, prolactin, also modulates sexual arousal by suppressing dopamine levels, which reduces feelings of frustration and antagonism.

While the baseline level of dopamine does not differ by sex, its release is greater in males than females in response to stimulation of desire. This difference helps to explain why college-age men think more often about sex, food, and sleep than women, although the gap between sexual cognitions is smaller than popularly assumed and likely to be overstated due to cultural expectations that suppress women's reporting (Fisher et al. 2012). Neurochemical processes involving dopamine also contribute to men's higher likelihood of alcoholism and other addictions (see Solinas et al. 2019). Dopamine mediates expectation and desire, increasing both tension – which contributes to motivation – and enjoyment. The second part of the equation, satisfaction of desire, is mediated mainly by endorphins. Drugs, pornography, and other shortcuts to the endorphin release brought on by activities such as exercise and sex eliminate the effort phase. Resulting neurochemical changes alter genetic expression in the brain, dysregulating the dopamine system and increasing susceptibility to addiction, which brings both sensitization (craving induced by ever smaller triggers) and tolerance (need for larger or more extreme doses to obtain pleasure).

Although aggression, violence, and addiction are more common among males, they are not exclusive to them. As with all behavioral traits, there is continuous variation within categories and overlap between them. Nearly all other cognitive and behavioral traits fall in the low range of *d* (for difference) values, which represent the difference between each group's average divided by its standard deviation. A difference greater than 0.8, whether positive or negative, is considered high. From 0.66 to 0.79 signifies moderately high; 0.36–0.65 moderate; 0.11–0.35 low; and below 0.10 little or no difference (Eliot 2010). Examples of traits with *d* values below 0.35 include higher trust and anxiety among women, greater self-esteem among boys and men; higher gregariousness and conscientiousness among women, more impulsiveness and openness to new experience among men (see Larsen and Buss 2009). Tests of math and verbal ability generate scores in the low range, whereas various tests of visuospatial ability result in *d* scores that range from no difference to moderate and, in some cases, high (Eliot 2010).

All such values are based on specific populations and cannot be assumed to represent all humankind. Most values are small in magnitude, generally about one-tenth the sex difference in height (2.0–2.6). If height were similar, it would be difficult to see a difference between women and men. Figure 5.2 summarizes some of the traits mentioned above and others discussed below.

With respect to academic and intellectual abilities, gender differences speak eloquently of the impact of cultural expectations, including the deviation of girls from socially valued studies in math and science. Girls and boys start out with equal math competence, as shown by research involving hundreds of children spanning in age from six months to eight years (Kersey et al. 2018). Skills were measured across several domains through tests of numerosity, verbal counting, and school-based math.[10] The results demonstrated a complete lack of gender differences in quantitative reasoning and math skills at all stages of early childhood math learning, clearly

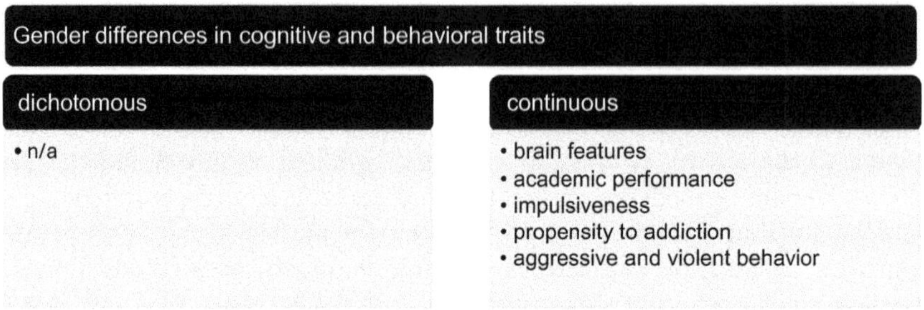


*Figure 5.2* Gender differences in cognitive and behavioral traits.

indicating that later divergences are due to social/cultural forces, not unequal inborn capacities.

The key role of essentialist beliefs in diverting girls away from math and science starting by middle school has been elucidated by a study in which about 300,000 respondents from 34 countries completed an Implicit Association Test in 17 languages through a virtual laboratory, the results of which were compared to test results from eighth-grade students in the same 34 countries who took the Trends in International Mathematics and Science Study (Nosek et al. 2009). Nearly three-quarters of respondents were more likely to associate math with males and liberal arts with females than the reverse. On the test, girls outperformed boys in countries with the weakest essentialist beliefs, whereas boys outperformed girls where gender stereotypes were more pronounced.

Tests administered every three years by Programme for International Student Assessment (PISA) to 15-year-olds in over 70 countries and economies affirm the association between math and science gaps and the degree of gender inequality in status, roles, and opportunities. Overall, girls and boys perform about the same on the science portion; there is a math gap in favor of boys; and there is a larger reading gap in favor of girls that is especially wide at the low end of scores (OECD 2023).[11] These differences mask the fact that girls do as well as or better than boys on the math portion of the test in half the countries and economies, and that the size of the math gap varies with the degree of gender inequality.[12] Girls outperform boys on the reading portion in all countries except in rare cases of equivalent scores and are more frequent among top performers. While boys continue to outnumber girls among top performers on the math test, the highest-achieving girls from countries such as Japan and Sweden outperform the highest-achieving boys from most other countries on both the math and science portions of the test.[13] In almost all countries, girls outperform boys at school in all subjects, while boys predominate among low performers. These disparities suggest the need for social and educational systems to address not only why the number of girls at the high end of the range on tests falls below expectation based on school performance, but also why boys are more likely to struggle at school.

While greater math anxiety and lower math self-confidence play a role in the lower performance of girls on standardized tests, an analysis of the variance across countries in the high-achievement disparity over five rounds of PISA tests points squarely to the impact of socioeconomic inequality (Breda et al. 2018). As few as three measures of socioeconomic inequality explained most of the variance: national-level income inequality; uneven learning opportunities; and, variable intergenerational mobility in education and earnings. Because boys enjoy higher status than girls across the board, higher levels of socioeconomic inequality translate into larger status gaps between genders and thereby to larger differences in test scores. Where there is more inequality, both boys and girls in disadvantaged groups perform worse at school, the effect is greater among girls, and the gap favoring girls in reading is narrower. Contrariwise, where there is less inequality, the top tier of performers on math tests includes higher numbers of girls as well as culturally and economically disadvantaged students.

In the US, the ratio of men to women among STEM majors in fields including chemistry, math, and biology is now close to 1:1. In physics, engineering, and computer science (PECS), it is 4:1. PECS fields are considered the hardest of the hard sciences, and consequently are the most male-associated, in line with the oppositions listed above. An analysis of STEM achievement in high school compared to intended and actual college majors indicates not only that more men than women pursue PECS along the entire achievement distribution but also that 90% of PECS majors in the lowest third of achievement are men (Cimpian et al. 2020). The data were gathered by the US Department of Education over seven years on thousands of students from ninth grade through the third year of college.[14] Of the students who went to four-year colleges, 23.8% of men and 5.5% of women majored in PECS. The lowest-achieving men (at or below the first percentile) were as likely (10%) to choose PECS as women at or above the 80th percentile. For every 1 woman in the bottom decile to choose PECS, there were 10 men; whereas in the top decile the proportion was only 1:2. Among students who majored in PECS but had not intended to, men under the first percentile of STEM achievement were as likely as women in the 99th percentile to choose a PECS major. Persistence in a PECS major was similar between men and women at high achievement levels but greater among men at lower levels.

While disparities in PECS majors did not occur between racial groupings (White, Asian, and Underrepresented Racial Minorities), gender gaps remained within them. In contrast, there were no notable gender or race disparities overall or along the achievement scale for other STEM fields. This contrast further highlights the role of masculine disciplinary cultures and social factors in keeping women out of PECS while encouraging low-achieving men to remain. The same male-favoring systems account for the low proportion of women in graduate PECS programs relative to the expected value based on their higher achievement level (20% women versus over 33% men). An analysis of student characteristics in relation to achievement and PECS leads the authors to conclude that interventions such as mentoring, peer networks, and coding camps work for high-achieving girls, but social and cultural changes are needed to attract and retain those with lower levels of achievement and to discourage the least well-suited boys from choosing and persisting in PECS fields. They argue that appealing to a broader range of girls would encourage other women down the line more effectively than focusing on top achievers.

A semester-long experiment with college students in a physics class illustrates how instructors can reduce the effect of male disciplinary cultures on women's learning and performance in PECS. Students in the intervention group completed writing exercises about personal values that were designed to increase feelings of self-worth. The activity boosted women's grades compared to the control group and equalized their improvement in knowledge to that of the men in the class. The largest effect occurred among women who entered the class with the strongest beliefs about innate male physics ability (Mikaye et al. 2010).

Even though essentialist gender beliefs do not match the evidence of grades and standardized tests, they persist in penalizing girls and women at school and in the workplace. People in senior positions including women scientists unconsciously favor male grant applicants, academic authors, and job candidates (see Moss-Racusin et al. 2012). The social role ideas discussed in Chapter 1 contribute to inequalities in compensation and career advancement in academia, which are related to gender disparities in publications and recognition through citations (see Casad et al. 2021). That women continue to be held to a higher standard by editors and reviewers, irrespective of their own gender, is evident from an analysis of 30,000 manuscripts published in four major economics journals. Published papers by women received 25% more citations than similar articles authored by men, and papers by women authors were less likely to receive a revise and resubmit verdict from editors (Card et al. 2020).

## Emotions

According to the oppositions listed above, masculinity lines up with intellect, femininity with emotion. However, the sex difference in aggression and violence, as well as experiments in which people are presented violent images or the threat of pain, suggests that, if anything, males tend to be more reactive and emotional than females (see Manstead 1992). The question is complicated by the fact that cultural beliefs shape the range of emotions women and men are expected to perceive and express (see Lutz and White 1986). For example, Geertz (1974) observed that in Bali and Java everyone regardless of sex is discouraged from expressing anger and other socially destructive emotions. The ideal is a smooth exterior state that troubles no one. In Euro-American culture, anger and frustration are tolerated in boys and men, sadness and apprehensiveness in girls and women. The former are strongly discouraged from crying, whereas the latter learn to suppress anger. Boys and men are expected to experience more positive emotions and to feel positive emotions in relation to themselves, whereas girls and women are expected to experience more negative emotions and to feel positive ones in relation to others. Here too, socioeconomic inequality plays a decisive role in a gender difference, for in more unequal societies there is a greater distance in emotional experience, expression, and literacy between women and men (Fischer and Manstead 2000).

Although contested (see Dobrin and Bashkow 2010), Margaret Mead's (1935) comparative analysis in Papua New Guinea found women and men expressing emotion in diverse ways that correlated with different socioeconomic and gender-role formulations. Among the Mundugumor, women and men both behaved in a hostile, competitive, uncaring manner. The peaceful Arapesh upheld "maternal" cooperative and nurturing attitudes as the ideal for both sexes. In the last group, the Tchambuli, women were industrious and spartan whereas men were vain, insecure, and dependent on women's attention.

Competitiveness, which is strongly coded as male in Euro-American society, depends on social context and especially the degree of gender inequality. In societies where women control property and economic decision-making, they tend to behave as or more assertively and competitively as men (see Brøgger and Gilmore 1997; Weismantel 2001). Experimental conditions reveal the same pattern of greater competitiveness with increased economic independence. Presented with an unfamiliar game of tossing balls into a bucket which they could play alone for one sum of money or against a competitor for a larger one, Tanzanian Maasai women were about half as likely as men play against a rival. Women of the matrilineal Khasi population in northeastern India were significantly more likely than men to choose to compete (Gneezy et al. 2009).

Cross-cultural variability in emotion concepts casts further doubt on sex differences in emotion, given the absence of a universal code of terms and meanings. A study of nearly 2,500 languages from 20 language families shows that, like single words that cover a range of colors, emotion concepts may signify multiple affective states along a spectrum (such as the

Persian word *amduh* for both grief and regret) (Jackson et al. 2019). Emotion concepts vary between language families, while there is graded similarity across geographical space. Languages typically classify emotions in terms of arousal and valence, albeit in variable ways: relatively individuocentric Western cultures define optimistic high-arousal states as optimal for emotional wellbeing, whereas comparatively collective-oriented Eastern cultures value somber, low-arousal states.[15] Similarly to emotion concepts, color terms vary across cultures and in complex relation to historical and environmental factors (Josserand et al. 2021). Events and visual fields likewise are viewed differently by people from Western versus Eastern societies: the former focus on individuals or objects within a frame, the latter on the overall context (Nisbett 2019). Even the senses are perceived in variable ways. While the English language prioritizes sight and hearing, a study of 20 unrelated languages found no evidence for a common human tendency to perceive these senses as more accessible to consciousness than others (Majid et al. 2018).

These differences in perception indicate the degree to which culture shapes even the most seemingly natural aspects of human experience. Emotional expression occurs within a local field of meaning that may associate certain feelings with a particular gender. Together with socioeconomic structures, emotion concepts and other cultural beliefs direct girls and boys toward activities, skills, and fields of study aligned with gender expectations. Disparities in brain features or average test scores are then taken as proof of inborn differences in behavior and cognition, despite the variability across societies and over periods of time too brief to reflect biological change. Comparison of averages collapses the continuous, overlapping ranges among girls and boys, which derive not from separate genetic essences but individual differences owing to variable genetic, environmental, and experiential processes.

## Sexualities, genders, gender expectations, and gender roles

To complete this chapter's anti-essentialist discussion of biological sex, physical sex differences, and behavior and cognition, the following pages present an overview of sexualities, sexual behavior, sex and gender identities, and culturally shaped gender roles.

The way biology textbooks explain sex, discussed in the last chapter, implies that its sole or primary purpose is reproduction. Human and animal behavior casts doubt on this assumption, for many species of mammals and birds regularly participate in same-sex sexual activity (see Bagemihl 1999). Bonobos engage in frequent female-female genital-to-genital ("G-G") rubbing and male-male "penile sword play" (de Waal and Lanting 1997). The activity calms everyone when the group comes upon an especially abundant food source or gathers at feeding time in captive populations, which could otherwise spur social tensions. Female newcomers ease their transition into a group and build relationships by engaging primarily in sexual activity with other females, and secondarily with males. In addition, opposite-sex sexual activity occurs when there is no possibility of pregnancy: bonobo females have sex with males while pregnant; their nearly always-visible sexual swellings invite sexual attention during periods distant from ovulation.

Sexual activity between individuals of the same sex has been observed in one-eighth of nonhuman primate species (59 of 491), including both chimpanzee species, both species of gorilla, and two of three species of orangutan (Coxshall et al. 2026). Analysis of 15 social, ecological, and life-history variables shows that the activity is more likely among species with complex social organization, greater sexual dimorphism in size, and ecological circumstances such as arid environments that bring increased predation pressure and food scarcity. As in the case of bonobos, same-sex sexual activity contributes to social bonding and trust. Alliances protect members, increase their access to food, and improve their position in the social hierarchy (see Chapter 7). These findings demonstrate the need for evolutionary theory to look beyond the immediate costs of non-reproductive sexual activity to the wider social and longer temporal horizon that impacts an individual's survival and reproductive success.

Human sexual behavior includes not only same-sex activity and opposite-sex interactions at times where there is no possibility of conception, but also purposeful blocking of conception through contraception, on-demand breastfeeding, post-partum restrictions on sexual activity, and other methods. Rarely do people have the maximum number of offspring possible, and this includes men, who allegedly are biologically programmed to spread their seed far and wide (see Chapter 6). Elite men in societies that permit polygyny are a rare exception. Historically, even where early weaning has multiplied the number of children women could have over their lifetimes (see Chapter 6), they and their male partners have often found ways to limit births (see Whitaker 2000).

Entrenched popular and scholarly assumptions about marriage maintain that it universally concerns romantic love, involves one woman and one man, has the purpose of generating children whose paternity is beyond doubt, and establishes parenthood that is simultaneously social and biological. The variability of marriage customs and family relationships around the world and within societies – including alternative kinship forms in mass societies – challenges these notions (see Levine 2008). Not only does marriage often involve strategic alliances arranged by families, but it may join multiple people and give children multiple mothers or fathers. Polygyny is or formerly was customary in many areas of Africa and the

Middle East, Madagascar, and parts of China, India, and the Americas. Polyandry, or multiple husbands, is or was practiced in Nepal, Tibet, northern China, parts of South Asia, some African and South American countries, Arctic Canada, and some Pacific Islands. Other forms of marriage include widows taking a wife to help manage the household; the prescribed union of a widow with her husband's brother; and the Nuer tradition of "ghost marriage" whereby a deceased unmarried or childless man's herds were used by a relative to pay bride price and ensure that the woman's offspring would be considered his descendants. Among the Samo of Burkina Faso, young women would spend years with a lover before entering an arranged marriage, bringing their first child with them as the husband's own child. The institution of partible paternity discussed in the next chapter gives multiple men lifelong responsibilities as biological fathers of the same child.

In addition to affecting the characteristics women and men seek in a spouse, noted above, gender inequality influences women and men's attitudes toward sexual behavior. Where there is a high degree of gender equality, women and men share similar beliefs about sex and sexual behavior and are equally likely to initiate sexual activity (see Weismantel 2001). In such societies, women are not penalized for having multiple partners or sex outside marriage. Through the institution of "walking marriage," practiced by matrilineal Na/Mosuo in China among other egalitarian societies, a woman signals her interest to a man during daily life (see Yong and Li 2022). She leaves her door open for him at night, and he leaves before morning. If they marry, the man continues to live with his relatives by birth and both spouses may continue to have multiple lovers. The fact that a child's father may remain unknown does not stop men from participating in raising their wives' children, calling into question the purported necessity of paternity certainty for male parental investment.

In contexts of marked gender inequality, in contrast, women's sexual behavior is tightly controlled and disallowed outside socially approved relationships, and divorce is difficult or impossible for women to obtain. The connection between controlled sexuality and cultural beliefs and legal systems that limit or preclude women's economic independence is evident in historical changes in divorce laws and restrictions on women's sexual behavior. To illustrate, until recent decades in the US, women's self-censorship made it difficult for researchers to reliably characterize their sexual beliefs, inclinations, and behavior, which fed into the traditional belief that cultural rules reflected natural differences. As women's economic independence has grown, their expressed attitudes have converged with men's. A corrected design for a study that decades earlier had shown women to be less interested in casual sex than men found that women had the same motivation in desired sexual pleasure and were equally open to an offer of casual sex, provided they perceived the prospective partner as possessing intelligence and sexual skill (Conley 2011).

The expectation of heterosexual desire flowing from the reproductive scope of the sex/gender binary conflicts with the regularity of same-sex sexual behavior in the living world. Among humans, a substantial proportion of every population chooses same-sex partners, and many more are forced by social and legal strictures to conceal their preference. Desire is not restricted to either same- or other-sex partners, may not be felt in any direction, and may change over time. Moreover, lived sex or gender identity is not dictated by chromosomes or reproductive anatomy. Many people identify as nonbinary or trans, although in the latter case identity and physical traits may be brought into alignment through hormonal and surgical interventions.[16]

While people whose sexual preference or sexual or gender identity falls outside the binary categories are often marginalized and discriminated against, in some societies they are recognized and may be honored. Historically, Native American two-spirits were individuals who could be either male-bodied or female-bodied and were free to choose idiosyncratic and changeable combinations of dress, behavior, and sexual partners (aside from other two-spirits) (see Lang 2016). Two-spirits might cook or make decorative objects and engage in warfare or hunting, for example. European observers assumed they were cross-dressing homosexual men and labeled them *berdaches*, the French rendition of a Persian word for a younger or passive partner in a homosexual relationship. In their own societies, two-spirits were held in high esteem and believed to have special access to the spirit world, which they harnessed for healing, divination, and storytelling. Many were elite, accomplished individuals. Their sexual or marriage partners were not considered homosexual. All North American indigenous societies either included two-spirits or lacked beliefs that would have excluded them. Contemporary societies likewise recognize two-spirits, although the sociocultural context has changed and brought negative impacts on their wellbeing and social position.

Male-bodied third gender categories exist in many parts of the world and, like two-spirits, are often associated with sacred powers (see Andaya 2018; Herdt 2020; Verma et al. 2023). Individuals are gendered as neither man nor woman, although in some cases they wear women's clothes. They represent a tri-equal gender, not an intermediate category within a binary system. Examples include *fa'afafine* in Samoa, *bissu* in Indonesia, *baklâ* in the Philippines, *kathoey* in Thailand, *hijra* in India and Bangladesh, *mahu* in Hawai'i, *khanith* in Oman, *travesti* in Brazil, and *muxe* among the Zapotec in Mexico. Despite social and, in some cases, legal recognition (as in the case of *hijras* in India), third-gender individuals may be subjected to social and economic marginalization; often, their main source of employment is sex work, domestic service, or the theater.

There are few recognized categories of female-bodied individuals who dress and/or behave in a neutral or masculine way, such as *tomboi* in southeast Asia. In many Muslim countries, girls in families without boys may dress as and live as boys, *bacha posh*, which permits them to work, attend school, move through public spaces, and run errands their sisters may not. At puberty these girls are expected to revert to a feminine identity. In contrast, northern Albanian "sworn virgins" live permanently as men and are thereby spared the toil and subservience assigned to women. Families without male children choose a girl to handle all the duties and activities of men, including blood feuds. "Sworn virgins" dress as men, drink and smoke, work outdoors, own and manage property, and host visitors. For some, the transition is so thorough and complete that they never go through puberty: their breasts do not develop, and they never menstruate (Young 2000).

Beyond the diversity sketched above, the activities considered appropriate for women and men vary around the world, often in mirror-image ways (see Ward and Edelstein 2015; Zeanah et al. 2026). There is no single, fixed gender pattern to the way foraging societies divide tasks, including building shelters, working wood and carving canoes, sewing, processing hides, sewing, cooking, pounding and decorating bark cloth, weaving baskets, and making nets for fishing and hunting.[17] Horticultural societies likewise divide tasks by gender in innumerable ways, including chopping trees; clearing or burning undergrowth; and planting, harvesting, and processing food. Moreover, contrary to the globalized symbolic opposition that construes men as producers and women as reproducers/caregivers, whether women are expected to or can realize the ideal of devoted motherhood depends on cultural, socioeconomic, and historical circumstances (see Scheper-Hughes 2023; Whitaker 2000). Contrariwise, men in small-scale societies bathe, carry, play with, and teach infants and children (see Chapter 6). For indigenous Andean men, being actively involved fathers and caring for relatives is considered masculine and bolsters their social status (Weismantel 2001), as is the case for men in many other cultural contexts, including ones known for *machismo* such as Mexico (see Gutmann 2007).

In short, a universal gender schema for behavioral, emotional, and cognitive traits does not exist. There are multiple sexes, sexualities, and sex and gender identities. Cultural systems set up different expectations, and individuals comply to varying degrees. As Margaret Mead (1935) concluded nearly a century ago:

> The material suggests that we may say that many, if not all, of the personality traits which we have called masculine or feminine are as lightly linked to sex, as are the clothing, the manners, and the form of head-dress that a society at a given period assigns to either sex.

## Notes

1 $n$ = 2,197 (versus the original 396). The difference was about twice as large ($d$ = 0.13, low range).
2 Stature is a more reliable measure than weight, given that lifestyle and environmental factors play a greater role in the latter. Based on a genomic analysis involving some 44,000 individuals, Yang et al. (2015) estimate that genes account for 60–70% of the variance in height but only 30–40% of the variance in body mass index in human populations.
3 While the authors note that genes associated with longevity may be more frequent in these relatively isolated populations, the decline in the number of centenarians born after the first decades of the 20th century points to the importance of nutrition and other lifestyle factors.
4 Sample range $n$ = 138–855.
5 Together more than 5,500 individuals.
6 $n$ = 157 women; $n$ = 106 men.
7 Hormone levels included estrogen, progesterone, luteinizing hormone, follicle-stimulating hormone, thyroid-stimulating hormone, and testosterone.
8 Visuospatial working memory test: repeat either in order or reverse order a sequence of nine colored blocks. Divided attention: 35 items shown in three phases, followed by quick identification of those corresponding to predefined visual and auditory cues by pressing a button with one or the other hand. Cognitive bias/control: rate pairs of items for shared binary qualities from 0 (none) to 5 (identical in all 5 ways) of red/blue, large/small, circle/square, one/two identical components, outlined or colored in.
9 For a history of the breast and other reproductive structures in visions of femininity, see King (2025).
10 Numerosity tests probed the ability to discriminate between unequal sets of objects in infancy and early childhood. Verbal counting among preschoolers involved memorizing and verbalizing sequences of numbers. Tests of early elementary school-based math included numerical knowledge, sets, word problems, and part-whole concepts.
11 On the 2022 test, the math gap was 9 points while the reading gap was 24 points among OECD countries.
12 On the 2022 test, girls scored higher in 17 countries and economies, there was no difference in 24, and girls scored lower in 40.
13 The ratio of girls to boys among those who score in the top decile on the math portion of the PISA test is relatively stable at around 7:10 across 35 OECD countries (Breda et al. 2018).
14 $n$ = 5,960.
15 The authors caution that similarities may be overstated due to issues related to translation and the use of a lingua franca during the research.
16 A review of 27 studies suggests that 871 per 100,000 (0.87%) identify as transgender, and the proportion of the population that undergoes surgical or hormonal therapy is about 100 times smaller at 9.2 per 100,000 (Collin et al. 2016). In the US, a Gallup poll of 14,000 adults determined that 9.3% of the population identifies as LGBTQ+, up from 3.5% in the first poll in 2012, including 1.3% who identify as transgender (Gallup.com 2025). The proportions increase as age group declines, with 23% of Gen Z respondents identifying as not heterosexual. Interviews with 900 individuals indicate the following proportions of non-straight identities, of which more than one could be chosen: 56% bisexual, 21% gay, 15% lesbian, 14% transgender, and 6% other (queer, pansexual, asexual).

17 Zeanah et al. (2026) correlate the breakdown of gendered tasks to the degree of dietary dependence on gathering, hunting, and fishing; gender contributions to the diet; and the amount of mobility.

## References


Andaya LY. 2018. The bissu. In Zamfira A, de Montlibert C, Radu D, editors. *Gender in focus*. Columbia University Press. pp. 64–87.

Bagemihl B. 1999. *Biological exuberance*. Macmillan.

Berry JW. 1966. Temne and Eskimo perceptual skills. *International Journal of Psychology* 1(3):207–229.

Beyene Y, Martin MC. 2001. Menopausal experiences and bone density of Mayan women in Yucatan, *Mexico. American Journal of Human Biology* 13(4):505–511.

Björkqvist K. 2018. Gender differences in aggression. *Current Opinion in Psychology* 19:39–42.

Block CL, Eroglu O, Mague SD, Smith CJ, Ceasrine AM, et al. 2022. Prenatal environmental stressors impair postnatal microglia function and adult behavior in males. *Cell Reports* 40(5):111161.

Bogardus T. 2025. *The nature of the sexes*. Taylor & Francis.

Breda T, Jouini E, Napp C. 2018. Societal inequalities amplify gender gaps in math. *Science* 359(6381):1219–1220.

Brøgger J, Gilmore DD. 1997. The matrifocal family in Iberia. *Ethnology* 1:13–30.

Card D, DellaVigna S, Funk P, Iriberri N. 2020. Are referees and editors in economics gender neutral? *The Quarterly Journal of Economics* 135(1):269–327.

Casad BJ, Franks JE, Garasky CE, Kittleman MM, Roesler AC, et al. 2021. Gender inequality in academia. *Journal of Neuroscience Research* 99(1):13–23.

Cimpian JR, Kim TH, McDermott ZT. 2020. Understanding persistent gender gaps in STEM. *Science* 368(6497):1317–1319.

Collin L, Reisner SL, Tangpricha V, Goodman M. 2016. Prevalence of transgender depends on the “case” definition. *The Journal of Sexual Medicine* 13(4):613–626.

Conley TD. 2011. Perceived proposer personality characteristics and gender differences in acceptance of casual sex offers. *Journal of Personality and Social Psychology* 100(2):309–329.

Coxshall C, Nesbit M, Hodge J, Savolainen V. 2026. Ecological and social pressures drive same-sex sexual behaviour in non-human primates. *Nature Ecology & Evolution* 12. https://doi.org/10.1038/s41559-025-02945-8

Davis JT, Robertson E, Lew-Levy S, Neldner K, Kapitany R, et al. 2021. Cultural components of sex differences in color preference. *Child Development* 92(4):1574–1589.

De Waal F, Lanting F. 1997. *Bonobo*. University of California Press.

Dettwyler KA. 1995. Beauty and the breast. In Stuart-Macadam P, Dettyler KA, editors. *Breastfeeding*. Aldine de Gruyter. pp. 167–216.

Dobrin LM, Bashkow I. 2010. “Arapesh warfare.” *American Anthropologist* 112(3):370–383.

Dunk RD, Malmquist SJ, Prescott KK, Ewell SN, Henning JA, et al. 2024. How do students critically evaluate outdated language that relates to gender in biology? *CBE—Life Sciences Education* 23(2):ar24.

Eastwick PW, Eagly AH, Glick P, Johannesen-Schmidt MC, Fiske ST, et al. 2006. Is traditional gender ideology associated with sex-typed mate preferences? *Sex Roles* 54(9):603–614.

Eaton SB, Pike MC, Short RV, Lee NC, Trussell J, et al. 1994. Women's reproductive cancers in evolutionary context. *The Quarterly Review of Biology* 69(3):353–367.
Edmonds, A. 2013. Can medicine be aesthetic? *Medical Anthropology Quarterly* 27(2):233–252.
Eliot L. 2010. *Pink brain, blue brain*. Simon and Schuster.
Elliott ZA. 2023. *Binary*. Paradox Press.
Ellis W, Bercovitch F, FitzGibbon S, Roe P, Wimmer J, et al. 2011. Koala bellows and their association with the spatial dynamics of free-ranging koalas. *Behavioral Ecology* 22(2):372–377.
Fausto-Sterling A. 2000. The five sexes, revisited. *The Sciences* 40(4):18–23.
Fischer AH, Manstead AS. 2000. The relation between gender and emotions in different cultures. In Fischer AH, editor. *Gender and emotion*. Cambridge University Press. pp. 71–94.
Fisher TD, Moore ZT, Pittenger MJ. 2012. Sex on the brain? *Journal of Sex Research* 49(1):69–77.
Fry DP. 1992. "Respect for the rights of others is peace." *American Anthropologist* 94(3):621–639.
Fry DP. 2006. *The human potential for peace*. Oxford University Press.
Fuentes A. 2025. *Sex is a spectrum*. Princeton University Press.
Gallup.com. 2025. LGBTQ+ identification in US rises to 9.3%. https://news.gallup.com/poll/656708/lgbtq-identification-rises.aspx
Geertz C. 1974. "From the native's point of view." *Bulletin of the American Academy of Arts and Sciences* 1:26–45.
Geller PL. 2009. Bodyscapes, biology, and heteronormativity. *American Anthropologist* 111(4):504–516.
Gneezy U, Leonard KL, List JA. 2009. Gender differences in competition. *Econometrica* 77(5):1637–1664.
Grant JA, Courtemanche J, Duerden EG, Duncan GH, Rainville P. 2010. Cortical thickness and pain sensitivity in Zen meditators. *Emotion* 10(1):43–53.
Gurven M, Hill K. 2009. Why do men hunt? *Current Anthropology* 50(1):51–74.
Gutmann MC. 2007. *The meanings of macho*. University of California Press.
Hegarty P, Smith A. 2023. Public understanding of intersex. *International Journal of Impotence Research* 35(1):72–77.
Heise LL, Kotsadam A. 2015. Cross-national and multilevel correlates of partner violence. *The Lancet Global Health* 3(6):e332–340.
Herdt G, editor. 2020. *Third sex, third gender*. Princeton University Press.
Jackson JC, Watts J, Henry TR, List JM, Forkel R, et al. 2019. Emotion semantics show both cultural variation and universal structure. *Science* 366(6472):1517–1522.
Jadva V, Hines M, Golombok S. 2010. Infants' preferences for toys, colors, and shapes. *Archives of Sexual Behavior* 39(6):1261–1273.
Joel D, Berman Z, Tavor I, Wexler N, Gaber O, et al. 2015. Sex beyond the genitalia. *Proceedings of the National Academy of Sciences* 112(50):15468–15473.
Jolly A. 2001 [1990]. *Lucy's legacy*. Harvard University Press.
Josserand M, Meeussen E, Majid A, Dediu D. 2021. Environment and culture shape both the colour lexicon and the genetics of colour perception. *Scientific Reports* 11(1):19095.
Kersey AJ, Braham EJ, Csumitta KD, Libertus ME, Cantlon JF. 2018. No intrinsic gender differences in children's earliest numerical abilities. *npj Science of Learning* 3(1):12. https://doi.org/10.1038/s41539-018-0028-7

Khan SI, Andrews KL, Jennings GL, Sampson AK, Chin-Dusting JP. 2019. Y chromosome, hypertension and cardiovascular disease. *International Journal of Molecular Sciences* 20(12):2892.

King H. 2025. *Immaculate forms*. Basic Books.

Laner MR, Russell JN. 1998. Desired characteristics of spouses and best friends. *Sociological Inquiry* 68(2):186–202.

Lang S. 2016. Native American men-women, lesbians, two-spirits. *Journal of Lesbian Studies* (3–4):299–323.

Larsen R, Buss D. 2009. *Personality psychology*. 4th ed. McGraw-Hill.

Leeners B, Kruger TH, Geraedts K, Tronci E, Mancini T, et al. 2017. Lack of associations between female hormone levels and visuospatial working memory, divided attention and cognitive bias across two consecutive menstrual cycles. *Frontiers in Behavioral Neuroscience* 11:120.

Levine NE. 2008. Alternative kinship, marriage, and reproduction. *Annual Review of Anthropology* 37(1):375–389.

Lock M, Kaufert P. 2001. Menopause, local biologies, and cultures of aging. *American Journal of Human Biology* 13(4):494–504.

Lutz C, White GM. 1986. The anthropology of emotions. *Annual Review of Anthropology* 1:405–436.

Maguire EA, Gadian DG, Johnsrude IS, Good CD, Ashburner J, et al. 2000. Navigation-related structural change in the hippocampi of taxi drivers. *Proceedings of the National Academy of Sciences* 97(8):4398–4403.

Majid A, Roberts SG, Cilissen L, Emmorey K, Nicodemus B, et al. 2018. Differential coding of perception in the world's languages. *Proceedings of the National Academy of Sciences* 115(45):11369–11376.

Manstead AS. 1992. Gender differences in emotion. In Gale A, Eysenck MW, editors. *Handbook of individual differences*. John Wiley & Sons. pp. 355–387.

Mead M. 1935. *Sex and temperament in three primitive societies*. Routledge and Kegan Paul.

Mehl MR, Vazire S, Ramírez-Esparza N, Slatcher RB, Pennebaker JW. 2007. Are women really more talkative than men? *Science* 317(5834):82.

Mikaye A, Kost-Smith LE, Finkelstein ND, Pollock SJ, Cohen GL, et al. 2010. Reducing the gender achievement gap in college science. *Science* 320(6008):1234–1237.

Morrison RE, Eckardt W, Colchero F, Vecellio V, Stoinski TS. 2021. Social groups buffer maternal loss in mountain gorillas. *Elife* 10:e62939.

Moss-Racusin CA, Dovidio JF, Brescoll VL, Graham MJ, Handelsman J. 2012. Science faculty's subtle gender biases favor male students. *Proceedings of the National Academy of Sciences* 109(41):16474–16479.

Nisbett RE. 2019. Cognition and perception. In Jing Q, Rosenzweig MR, d'Ydewalle G, Zhang H, Chen H-C, et al, editors. *Progress in psychological science around the world. Volume 2: Social and applied issues*. Routledge. pp. 209–228.

Nosek BA, Smyth FL, Sriram N, Lindner NM, Devos T, et al. 2009. National differences in gender–science stereotypes predict national sex differences in science and math achievement. *Proceedings of the National Academy of Sciences* 106(26):10593–10597.

OECD. 2023. *PISA 2022 results (Volume I)*. OECD Publishing. https://doi.org/10.1787/53f23881-en

Pes GM, Tolu F, Dore MP, Sechi GP, Errigo A, et al. 2015. Male longevity in Sardinia, a review of historical sources supporting a causal link with dietary factors. *European Journal of Clinical Nutrition* 69(4):411–418.

Peters RM, Hackeman E, Goldreich D. 2009. Diminutive digits discern delicate details. *Journal of Neuroscience* 29(50):15756–15761.
Rosenbaum S, Vigilant L, Kuzawa CW, Stoinski TS. 2018. Caring for infants is associated with increased reproductive success for male mountain gorillas. *Scientific Reports* 8(1):15223.
Sax L. 2002. How common is intersex? *Journal of Sex Research* 39(3):174–178.
Scheper-Hughes N. 2023. *Death without weeping*. University of California Press.
Seddig D, Davidov E. 2018. Values, attitudes toward interpersonal violence, and interpersonal violent behavior. *Frontiers in Psychology* 9:604. https://doi.org/10.3389/fpsyg.2018.00604
Shapin S. 2022. Hard science, soft science. *History of Science* 60(3):287–328.
Sharma S, Gibbons A, Saphire EO. 2025. Sex differences in tissue-specific immunity and immunology. *Science* 389(6760):599–603.
Solinas M, Belujon P, Fernagut PO, Jaber M, Thiriet N. 2019. Dopamine and addiction. *Journal of Neural Transmission* 126(4):481–516.
Sorge RE, Strath LJ. 2018. Sex differences in pain responses. *Current Opinion in Physiology* 1(6):75–81.
Staniloiu A, Markowitsch H. 2012. Gender differences in violence and aggression–a neurobiological perspective. *Procedia-Social and Behavioral Sciences* 33:1032–1036.
Tidwell CA, Danvers AF, Pfeifer VA, Abel DB, Alisic E, et al. 2025. Are women really (not) more talkative than men? *Journal of Personality and Social Psychology* 128(2):367–391.
Tombak KJ, Hex SB, Rubenstein DI. 2024. New estimates indicate that males are not larger than females in most mammal species. *Nature Communications* 15(1):1872.
Verma SK, Sharad S, Singh T, Bharti P. 2023. Living with alternatives. *Culture & Psychology* 29(4):840–863.
Walter KV, Conroy-Beam D, Buss DM, Asao K, Sorokowska A, et al. 2020. Sex differences in mate preferences across 45 countries. *Psychological Science* 31(4):408–423.
Ward MC, Edelstein MD. 2015. *A world full of women*. Routledge.
Weismantel, Mary 2001. *Cholas and pishtacos*. University of Chicago Press.
Whitaker ED. 2000. *Measuring mamma's milk*. University of Michigan Press.
Wolfe LD, Gray JP. 1982. Latitude and intersocietal variation of human sexual dimorphism of stature. *Human Ecology* 10(3):409–416.
Yamauchi Y, Riel JM, Ruthig VA, Ortega EA, Mitchell MJ, et al. 2016. Two genes substitute for the mouse Y chromosome for spermatogenesis and reproduction. *Science* 351(6272):514–516.
Yang J, Bakshi A, Zhu Z, Hemani G, Vinkhuyzen AA, et al. 2015. Genetic variance estimation with imputed variants finds negligible missing heritability for human height and body mass index. *Nature Genetics* 47(10):1114–1120.
Yaskin VA. 2011 Seasonal changes in hippocampus size and spatial behavior in mammals and birds. *Biology Bulletin Reviews* 1(3):279–288.
Yong JC, Li NP. 2022. Elucidating evolutionary principles with the traditional Mosuo. *Culture and Evolution* 19(1):22–40.
Young A. 2000. *Women who become men*. Berg.
Young LC, Zaun BJ, VanderWerf EA. 2008. Successful same-sex pairing in Laysan albatross. *Biology Letters* 4(4):323–325.
Zeanah DW, Elston RG, Codding BF. 2026. Gender convergent labor and technological change at the Pleistocene to Holocene Transition in the Great Basin. *Journal of Anthropological Archaeology* 81:101746.

# 6 In search of ultimate causes

## Paleonarratives about sex and gender

The previous chapter raised a series of challenges to essentialist beliefs about sex-specific cognition, including the spectrum of sexes; brain plasticity; the variable and overlapping ranges of cognitive and behavioral traits; and cross-cultural variability in gender expectations and roles. While Chapter 3 exposed the evolutionary fallacies of scientific racism, this chapter confronts the notion that alleged sex differences in thought and behavior are rooted in a particular evolutionary scenario in which females and males carried out separate, non-overlapping tasks. Engagement in childrearing and gathering versus toolmaking and hunting allegedly inscribed human gender differences upon the genome forever, leaving women dependent, nurturing, verbal, and emotional. Men, in contrast, purportedly received the gifts of rationality, visuospatial ability, independence, and tactical reasoning. Not only do these arrays of characteristics fail to match the evidence presented in the last chapter, but they are unsupported by current knowledge of human origins and evolution, as shown below.

### Visions of prehistory

According to the traditional paleonarrative echoed by textbook explanations of human evolution, ancestral males hunted large animals to provision themselves and their dependents. Females were too burdened by hordes of offspring to do much other than pick fruit or berries close to the cave. The core of foraging groups was sets of related males who directed their competitive instincts outward by collaborating with each other for warfare and raiding. Stealing food and females from other groups supposedly increased the number and survival of their collective offspring, sending more of the patriline's genes into the next generation.

This series of ideas is neither logical nor supported by evidence. Females of other mammalian species procure their own food, including while they are pregnant or lactating. Human metabolism cannot handle an all-meat diet, while hunting large animals with tusks, antlers, and huge canines is

DOI: 10.4324/9781003639923-6

impracticable for single individuals or small groups carrying nothing but clubs or spears, as was the case until recently in evolutionary time. In a vast landscape with sparse inhabitants and abundant food resources, mobile foragers are highly unlikely to have been territorial or engaged in raiding and violent conflict. Finally, if males needed to protect females from warring parties of other males, they could not have afforded to leave their mates and children at home alone.

Paleolithic humans were tall, strong, and fit from physical activity and the varied diet associated with nomadic foraging. Females and males both had powerful muscles and robust bones, like top athletes today, as indicated by muscle insertion sites, joint surfaces, and broad fingers. At least until the Late Pleistocene, both their skeletons carried the signs of trauma and wear, indicating that females engaged in the same activities as males and were not sheltered from danger or harm. Evidence of warfare appears only recently, during the Holocene beginning 11,700 BP, and does not suggest that warfare between foraging groups was common. As shown in Chapter 7, extensive farming tended to bring increased conflict within and between populations, although not always.

The following discussion challenges the conventional narrative through a combination of ethnographic information on recent and contemporary foraging populations, archeological and paleontological data, and human biological traits related to endurance exercise. There are limitations to each of these sources. For starters, while nomadic foraging tends to be associated with certain social and cultural features, existing foragers do not represent a timeless human condition. Like all types of society, their way of life, degree of egalitarian social organization, and belief systems vary and change over time, as do the ecological environments they occupy. All recent and contemporary foragers have been restricted by states and commercial interests to marginal lands that are unrepresentative in terms of the population sizes they can support. Moreover, virtually all groups live near and engage in trade with farmers and/or herders. Even in the 1960s and 1970s, when anthropologists began to study these societies, it was rare for most of the food to come from foraging. In short, no pure or unchanging version of a foraging society has ever existed, and the socioecological circumstances of current groups are unlike those of prehistoric foragers.

Another limitation involves the assumption that men's activities are more interesting and important than women's. This bias runs through much of the literature on foraging societies, including the 1968 *Man the Hunter* volume edited by Richard B. Lee and Irven Devore, which followed upon a conference of the same title they organized in 1966. Researchers subsequently responded with publications detailing women's contributions to caloric intakes, such as Frances Dahlberg's 1981 edited volume, *Woman the Gatherer*. While the latter forced a reckoning with women's productive roles in foraging societies, the emphasis on gathering further obscured the overlap between women and men's food-procurement activities and thereby upheld the man/hunter, woman/gatherer opposition. For instance, one of the contributors to Dahlberg's book defined Tasmanian women's hunting of seals as "gathering" (in Sterling 2015). Nearly a half-century later, in a refutation of two articles presenting an argument for female hunting throughout human evolution, Martin et al. (2024) classified women's small-game hunting as "foraging."

Traditional gender beliefs have likewise shaped the way archeologists present and interpret data on prehistoric societies. For instance, stone and bone tools are presumed to have been manufactured by males. Tools for hunting are seen as requiring more skill and intelligence than ones used for processing food, a task symbolically associated with femininity. These biases are evident in visual and narrative presentations in textbooks and scholarly publications that give more prominence to hunting than non-hunting tools.

Cultural assumptions about sex and gender also influence the analysis of skeletal remains. In line with the presumed universality of sex as a master category, archeology textbooks present sex identification as the first topic in skeletal analysis, while research and media reports on prehistoric burials highlight sex before other characteristics. Rarely does either one note the chance of misidentification due to the limitations of pelvic measurements, especially on older-age individuals and nulliparous females, or the tendency to assume male sex where grave goods include hunting equipment or weapons (see Anderson et al. 2023).[1] Not only may sex be misinterpreted, but contemporary gender beliefs are often projected onto the lives of individuals who lived hundreds or thousands of years ago. For instance, marks of trauma or wear on male skeletons are more likely to be attributed to work, whereas those on females are interpreted in terms of reproduction (see Geller 2009). Multiple burials or footprints that seem to conform to heteronormative assumptions about marriage and family receive more attention than those which do not. Many must be reclassified upon further analysis, such as a presumed mother and father with two small children buried by the ash at Pompei that have turned out to be four unrelated males (Pilli et al. 2024).

Lastly, scientific studies on sex differences in human biological traits related to endurance activity are limited by the convention of comparing two averages, which reinforces the idea of a sex binary instead of a spectrum. As in the case of pelvic measurements, average values conceal the variation within categories. Whereas essentialism capitalizes on average values as expressive of absolute differences between categories, the overlapping distribution of data points speaks to shared characteristics and variability within both groups, as seen in the last chapter and discussed further below.

## Reproduction, group composition, kinship

Birth spacing and family size among ethnographic foraging societies undermine the popular image of prehistoric females surrounded by babies and small children (see Eaton et al. 1994). Babies typically are born three to five years apart, thanks largely to on-demand breastfeeding day and night for two to four years. This allows women to carry children in a sling until they are grown enough to walk on their own, and therefore to continue procuring food without interruption. As noted in Chapter 5, women ovulate from around age 18 to their early to mid-40s, when not pregnant or breastfeeding. During their reproductive years they have about five children, just over half of whom, on average, survive to adulthood.

For women in agricultural societies, earlier first menstruation and later menopause expand the reproductive phase of life by a decade or more. The availability of soft foodstuffs allows mothers to begin substituting breastmilk with other foods early on, which brings down the age at weaning and returns women to fertility within months of childbirth. As a result, women may have a dozen or more children over their lifetime. These differences explain why population began to grow rapidly with the Neolithic transition to agriculture, even though infant mortality rates were higher for agriculturalists than foragers.

Men in foraging societies also close their reproductive years at an earlier age than men in agricultural societies. To illustrate, among the Ju/'hoansi (!Kung), Hadza, and Agta, men have their last child in their mid-50s. In the mixed foraging-horticultural societies of the Yanomami, Tsimané, and Aché, they stop in their early to mid-60s. Among Turkana pastoralists and Gambian farmers, men continue reproducing into their 70s. While the maximum age at last reproduction across all these groups varies from 45 to 58 for women, the range for men is much wider, from 54 to 78 (Vinicius et al. 2014).[2]

The assumption that prehistoric groups of foragers were organized around sets of patrilineal kin likewise does not match the evidence. Not only are patrilineal groups uncommon, but few of the people who compose a residential group are closely related. The reason is that it is not just women who leave their natal group upon marriage. An ethnographic analysis of kin relationships among more than 5,000 adult members of 32 foraging societies shows that only 7% of adult group members are primary kin: on average, 1.8 individuals out of a total of 28.2 (Hill et al. 2011). The number of parents living with daughters versus sons is equal in 19 societies, greater for daughters in 6, and higher for sons in 7. The number of parents of women in a group is about the same as the number of parents of men (0.39 and 0.44). There is a strong preference for siblings to live together, including brothers with sisters, but it is more common for brothers than sisters to co-reside. The average number of brothers living with a man's group is larger than the number of sisters (0.66 versus 0.45), whereas for a woman the number is the same (0.42 versus 0.40). Although the total number of close co-residing kin is higher for men than women (1.93 versus 1.67), people of both sexes co-reside with few close relatives. This is especially the case for larger groups, in which the proportion of unrelated individuals is greater.

Low relatedness is likely to have been the case in the deep past as well, as shown by a model that compares what happens if the preference for co-residing with siblings is combined with either one sex having a say over who lives in the group or women and men having equal say (Dyble et al. 2015). In both cases, sibling preference results in more related individuals within groups than would occur if people joined groups at random. There is little or no difference in the number of biological relatives between wives and husbands. These results closely match the proportions of relatives and nonrelatives among the Aché and Ju/'hoansi determined through genetic analysis as part of the above study, as well as two additional foraging societies, the Agta and Aka (Mbendjele). The model predicts that about one-half of group members are the self, the spouse, and their primary and distant kin. The other half is composed of people unrelated to either spouse plus distant kin through marriage. Across the four groups, this last category is slightly smaller than one-half among the Agta, and substantially larger than one-half among the Ju/'hoansi. The proportion of non-relatives of either spouse ranges from less than one-eighth to more than one-fourth of the total.

The non-egalitarian model predicts that virtually no one will be unrelated in some way to one or the other spouse. About three-quarters will be the spouse and their primary and distant kin, while one-fourth will be distantly related by marriage. This prediction is close to the distribution of actual kinship relationships among Paranan farmers in the Philippines: under 5% are unrelated individuals; about two-thirds are primary or distant kin of the two spouses; and, about one-third are distant kin through marriage. Men live with more of their relatives than do women. The non-egalitarian model shows that sex inequality in residential choices causes high degrees of relatedness in agricultural societies, whereas the egalitarian model confirms that equality brings low relatedness within foraging groups.

The patrilineal sets of male relatives at the center of paleonarratives about warring tribes cannot form where residential groups contain few related individuals. Two ethnographic samples of 39 and 21 foraging societies, with some overlap, attest to the rarity of patrilineal kin groups (Fry 2006). In the first of these, 72% of societies lack such groups altogether, and only 26% practice patrilocal residence. In the second sample, 86%

lack patrilineal kin groups, and only 10% practice patrilocal residence. Kinship ties are recognized bilaterally, along both maternal and paternal lines, in 59% of the first sample and 71% of the second. These data concur with a generalized pattern by which unilateral postmarital residence or kinship organization is rare among foraging societies, especially where group size is small, as it was until the Late Pleistocene (see below and Lacy and Ocobock 2024). The patriarchal-patrilineal system that removes women from their birth families upon marriage is characteristic of agricultural societies and consequently recent.

Bilocal residence and low relatedness among group members generate complex networks of far-reaching genetic and in-law connections across residential groups. In turn, cross-cutting kinship ties and long-term exchange and ceremonial relationships favor the peaceful relationships characteristic of foraging societies. The dispersal of relatives is a well-established pattern, as indicated by a group burial in western Russia around 34,000 BP in which the four males – one previously identified as female – were not related to the first, second, or third degree (Sikora et al. 2017). Low relatedness and wide-ranging linkages among residential groups are likely to have been crucial to the evolution of human traits of high cooperativeness, shared parenting, and egalitarian attitudes and social arrangements (see Hill et al. 2011, 2014). In other words, cooperation does not stem from the efforts of related, co-resident males to conduct warfare for purposes of sending more shared genes into the next generation than they could have individually through pure self-interest.

## Sharing, caregiving, parenting

The actual rather than imagined organization of foraging societies described above indicates that humankind's exceptional capacity for cooperative behavior emerged through mechanisms beyond kin selection. For instance, food sharing is a universal behavior across foraging societies, is not limited to relatives, and was a key factor in the evolutionary success of the human species (Jaeggi and Gurven 2013). Sharing buffers risk and facilitates the creation of alliances and reproductive partnerships (see Townsend et al. 2021). In addition to everyday food sharing, groups share when they meet for seasonal gatherings, to access water or food sources, or whenever one group comes across an especially abundant food source.

Caregiving of the injured, ill, and disabled is an ancient form of sharing evidenced by burials of individuals with congenital defects, diseases, and injuries whose age indicates many years of attention from others. Far from being shunned, these individuals were buried with their communities. Sometimes, they were given extra honors as is common for individuals with physical anomalies in many non-industrialized societies. Two cases described by Schrenk and Tilley (2018) illustrate collective care. The first, a male who died in his mid-twenties around 4,000 BP in northern Vietnam, had lived for a decade with lower-body paralysis and highly limited use of his arms. His foraging community must have provided food, moved him when they changed residence, and managed the serious medical issues associated with paralysis. The second, an 18-year-old female in a farming village near today's Dubai 4,000 to 2,000 BP, was left with lower-body paralysis after contracting poliomyelitis. Her community must have nursed her through the months-long acute phase of sickness before providing her long-term care such as meals and medical treatment. Marks of heavy use on the upper limbs suggest that she dragged herself around or engaged in work that could be done while seated. Hers is the only intact skeleton in a collective grave full of mixed bones, possibly a sign of reverence.

Sharing in foraging societies also takes the form of mutual protection. In the deep past, protection was particularly needed against dangerous predators, including felines, crocodiles, huge snakes and enormous raptors, and dozens of species of hyenas, many of them much larger than hyenas today (Hart and Sussman 2005). Safety in numbers is a main reason all diurnal primate species live in stable residential groups.

Assisted childbirth is likely to be as old as humankind, for upright posture brought changes to the shape and rigidity of the pelvis that made childbirth difficult to manage alone (Trevathan 2015). In humans, the pelvic opening runs side to side at the top but front to back at the bottom, while the fetus's shoulders lie perpendicular to the widest plane of its head. The fetus consequently must make a series of rotations to exit the birth canal, at which point it faces behind the mother. A mother giving birth alone cannot easily wipe away fluids that may block respiration, release an umbilical cord wrapped around the neck as occurs in up to one-third of births, or guide the newborn out and to the breast without risking damage to its neck.

While the mechanics of childbirth call for assistance to mothers, the fact that human babies are born at a very early stage of neurological development calls for shared child care, or alloparenting.[3] Comparing across species, high levels of alloparenting are typical of those which carry infants over long distances, making humans more similar to langurs despite being more closely related phylogenetically to chimpanzees (Hrdy 2011). In human societies, assistance to children and their mothers begins before birth and continues for many years. Throughout the world, women are given extra food and a lighter workload during pregnancy and the postpartum period. The custom of partible paternity institutionalizes extra assistance to mothers through the establishment of kinship ties between the child and multiple fathers (see Beckerman and Valentine 2002). Societies with partible paternity are found in many places, including South America,

India, New Guinea, and the Philippines. To illustrate, Barí women in Venezuela are encouraged to take extra lovers during pregnancy to spare their husbands the exhaustion of feeding the fetus with "bathings" of semen. The additional fathers give the woman gifts of nourishing foods and remain biological fathers to the child for life, which favors both maternal health and child survival (Walker et al. 2010).

Ways that ancestral males could assist females include giving gifts of food to pregnant and lactating females, leaving closer food sources to them during pregnancy and lactation, and carrying infants unable to grasp with their feet the way nonhuman primates do. Just carrying an infant spares the mother an energy expense equal to lactation. It reduces suckling frequency and returns the mother to fertility sooner. Long ago, caregiving by fathers and others permitted females to wean their babies much earlier than the six years predicted by comparisons with other primates, while a more favorable energy balance improved maternal health and consequently children's chances of survival (Gettler 2010).[4] Paternal care allowed humans to outbreed other species, but it was not the same as provisioning, much less permanent provisioning of females.

Hormonal changes associated with childbirth and parenting confirm the antiquity of intensive male involvement with offspring (Gettler et al. 2012; Rilling and Young 2014; Rilling et al. 2025). While testosterone in single men favors their pursuit of a romantic partner, its level drops upon the birth of a child. Reduced testosterone is associated with increased empathy and less erratic and antisocial behavior. Sleeping with and providing daily care to babies brings further reductions in testosterone. Across individuals, reduced nurturance and responsiveness to infant needs are associated with lower levels of estradiol and higher levels of testosterone as well as vasopressin, whose many effects on social behavior include aggression and anxiety-like behaviors in response to stress, especially in males and particularly toward the same sex. Lower levels of testosterone in fathers are associated with better growth and survival rates in offspring. In addition, childbirth results in a rise in oxytocin and prolactin levels in fathers, not just mothers. Oxytocin favors relationships through feelings of other-orientation, tenderness, and trust. Prolactin promotes nurturing behavior and dampens sexual arousal, as noted in the last chapter. These and other hormonal changes open the door to new mental processes and reduced self-orientation, with intergenerational effects on romantic and parenting relationships.

Multiple caregivers bathe, comfort, and play with babies in foraging and other small-scale societies. Lactating women may breastfeed each other's children, while men and nonlactating women may put babies to the breast to calm them (see Hewlett 2017). To quantify the extent of alloparenting, 10 infants in 7 camps of seminomadic Efe foragers in the Democratic

Republic of the Congo were observed for two to four hours at 3, 7, and 18 weeks of age (Tronick et al. 1987). Over the course of the observations, an average of 14.2 people cared for each infant. At 3 weeks, the infants spent 39% of their time in physical contact with people other than their mothers; at 18 weeks, the amount of time was 60%. Babies were transferred between people 3.7 times per hour at 3 weeks; 5.6 times at 7 weeks, and 8.3 times at 18 weeks. Eight of the 10 infants were nursed by women other than their mothers, including women who were not lactating. Caregivers responded within 10 seconds of a baby's fussing in over 85% of instances during the first 7 weeks and over 75% at 18 weeks.

Alloparenting benefits children's health and allows parents to collect more food than they would otherwise. Maternal grandmothers have an especially positive impact on child survival across a wide range of societies (Sear and Mace 2008). This may explain the evolution of menopause (Hawkes et al. 1998) although, as noted above, forager men also experience a post-reproductive phase of life. Contrary to common misconceptions about old people being considered a burden or abandoned to their fate, aging women and men are valued for the stability, knowledge, and guidance they bring to the group (see Amoss and Harrell 1981).

Alloparenting, caregiving, and food sharing exemplify the interdependencies characteristic of foraging societies, which are likely to have been even stronger in the past when groups were smaller. Through most of the Pleistocene, groups consisted of about 30 individuals, although they could be as large as 110, and Neandertal groups were smaller yet (Lacy and Ocobock 2024). In the Late Pleistocene and Holocene, group size increased to 30–50 and up to 150. These larger sizes allow for specialization, including divisions of labor by gender that often include greater male involvement in large-game hunting (see below). While this shift could be expected to generate inequality, foragers perceive status competition as threatening to peace and individual autonomy and maintain social rules and cultural beliefs that prohibit hunters from using highly valued meat strategically. The stated purpose may be different – such as protecting hunters from ill fortune or demonstrating respect for nature or spirits – but the effect is to prevent individuals from gaining prestige and power over others.

Wiessner (1996) divides leveling mechanisms related to hunting into three main categories, applied singly or in combination and summarized in Figure 6.1: the expectation of modesty and self-effacement on the part of hunters when they return to camp or while distributing portions; traditions that assign the distribution of meat to someone else or to multiple others; and, exclusion of the concept of obligation, unless it is to construe sharing as a debt the hunter owes to the group. Specific examples include rules against hunters eating meat from animals they have killed or carrying them back to camp. Meat distribution may be assigned to the owner of the

**Leveling mechanisms related to hunting**

- systems that separate hunting from meat distribution
- exclusion of the concept of obligation, except for the obligation to share
- modesty and self-effacement expected of hunters

*Figure 6.1* Categories of leveling mechanisms.

arrow that killed the animal, which induces hunters to use borrowed arrows and avoid the responsibility of fairly allocating portions. Meat is always divided up, whether first among the hunters who then split it with their families, or back at camp where it is distributed to everyone.

In sum, the popular image of greedy males competing for control over resources needed by helpless females does not conform to the evidence that small groups of mostly unrelated foragers are likely to have shared food, collectively taken care of the young, and remained relatively egalitarian throughout human evolution. Exceptions involving larger groups and ones with access to especially abundant resources that have tended to be less egalitarian are discussed in the next chapter.

## Stone, wood, fire

According to the essentialist ideas about gender-divergent attributes described in the last chapter, males have exceptional visuospatial abilities due to an evolutionary history of hunting and associated stone-based tool-making. This scenario is at odds with the way ancestral humans procured food, processed plant and animal matter, and manufactured tools.

As noted in Chapter 3, the arid-savannah backdrop of conventional paleonarratives misrepresents the great variety of environments inhabited by human ancestors and the vast climatic shifts that altered the type and availability of foods they could access. Over time, the diet increasingly incorporated protein from terrestrial animals, birds, and inland and marine aquatic species. Yet, the increased protein could not have come from large-game hunting until effective implements were invented (see below). Until then, human ancestors gathered and scavenged. Their physical characteristics matched a life of long distance walking in daytime. Most predator species must hunt at night to avoid overheating, whereas humans have a highly efficient cooling system that includes sweat glands distributed all over the body plus small hair follicles and consequently a thin coat that further favors evaporative heat loss. Some hairiness was retained, such as protective eyebrows and lashes that shade the eyes, and head hair that reduces sun exposure and maintains a more constant brain temperature.

Hair in the genital and armpit areas may contribute to pheromonal signaling, while facial hair in males may reflect sexual selection (Cohn 1998).

Until hunting implements came along, and afterward as well, ancestral humans obtained meat from animals that died from natural causes, or by chasing away carnivores such as hyenas (see Hart and Sussman 2005). Marks made by blades and hand axes appear on the same bones as teeth marks from other animals. That Pleistocene foragers scavenged is evident from the mammoths and other animals consumed near sources of water or salt where they died. Had they hunted the animals, they would not have hauled back the 200-pound skulls that have been found among the butchered bones. Contemporary foragers including the Ju/'hoansi and Hadza continue to obtain a significant portion of their meat by scavenging.

The timing of the emergence of hunting implements varied across regions (see Adovasio et al. 2016; Ben-Dor and Barkai 2023; Hart and Sussman 2005). Roughly, the ability to hunt began 400,000 to 300,000 BP with throwing sticks and long spears, which may have been used for hunting or to chase predators away from their kills. Stone and bone tips found in southern Africa date to 75,000 to 60,000 BP, with poisoned tips appearing 60,000 BP; sticky materials for attaching components of tools such as pitch, beeswax, and tree resin, to 44,000 BP. The humans who arrived in Europe 45,000 BP brought projectile points. Later came the levered spear thrower (40,000 to 30,000 BP), knotted hunting nets (28,000 BP), fishhooks (at least 23,000 BP), and bows and arrows (at least 20,000 BP). Dogs were domesticated from wolves 20,000–30,000 or more years ago and had subdivided into several lineages by the Early Holocene, with the oldest genetically confirmed specimens dating to 15,800 BP (Bergström et al. 2020; Marsh et al. 2026). Figure 6.2 arranges these dates in a timeline.

The hunting-related innovations of the past 30,000 years reduced the need for upper body strength relative to throwing a spear. In particular, the bow and spear thrower greatly reduced the difference that strength makes to the velocity of a projectile. These tools broadened the pool of prey animals and could be used by females even while pregnant or carrying a baby. Net hunting was and remains a particularly important technique that involves the entire group in rounding up animals both large (bison, elk, bighorn sheep, as among Native American societies) and small (fox, hare, birds, small deer in Africa and historically in Europe). Nets may have been the most prized equipment, as they are among contemporary Mbuti foragers in the Central African Republic. Hunting with nets and other tools became so effective beginning 30,000 BP that, in Europe, it may have contributed to the disappearance of herds of herbivores as well as large predator animals such as saber-toothed tigers. However, foragers are not known for overharvesting. An analysis of contemporary Amazonian groups,

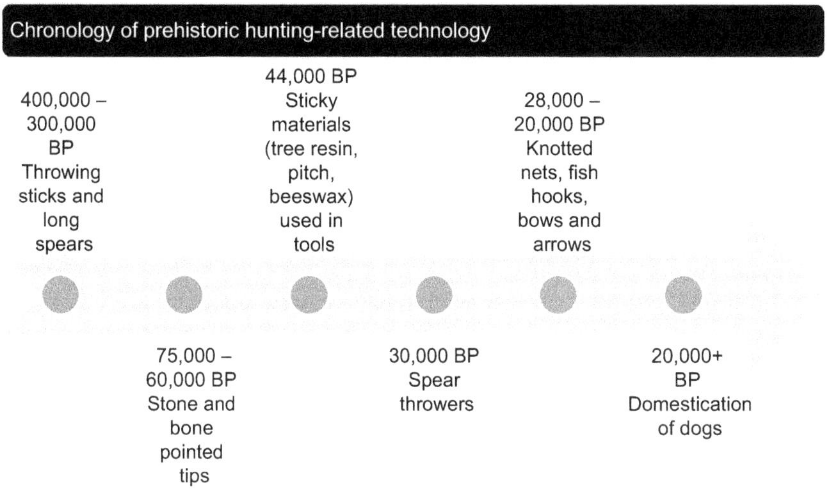


*Figure 6.2* Timeline of prehistoric hunting-related technology.

combined with computer simulations, shows that forager hunting practices do not deplete game – including large mammals with low reproductive rates – provided two conditions hold: cultural beliefs discourage excess, and firearms are not available (Shepard et al. 2012).

As shown below, the ethnographic literature contains abundant descriptions of hunting on the part of women. Archeological and paleontological data lend further support to the idea that Pleistocene females, not just males, engaged in hunting. Lacy and Ocobock (2024) report that, across Eurasia (albeit with some regional variations), no disparities have been found in the way Late Pleistocene females and males were treated after death, with no difference in grave goods or the frequency of individual burial. In addition, there are no differences in dental wear marks – presumably due to processing hides – between Neandertal females and males. The owner of an Upper Paleolithic pendant made of a deer tooth has been genetically identified as female, suggesting experience with hunting.

The assumption that "Stone Age" hunter/warrior males made all the tools in the archeological record conflicts with the fact that contemporary foragers produce the tools they use personally (Lacy and Ocobock 2024). In small, unspecialized groups of prehistoric foragers, everyone would have needed to make their own tools for collecting and processing materials for clothing, shelter, and food, in addition to protecting themselves from predators. Tools would have included digging sticks, scrapers and awls, grinding stones, implements for piercing hides and tough plant matter, clubs, spears, hand axes, and hunting implements.

The female stone-working tradition maintained by Konso women in southern Ethiopia today casts doubt on the traditional equation of males with lithic technology (Arthur 2010). Konso women learn to knap flaked stone tools for working hides through apprenticeship and long practice, using stone they procure from distant sources. Making stone implements is considered women's work, so much so that men refuse to engage in it, indicating that the task has long been associated with femininity.

In addition to its symbolic association with masculinity, lithic technology's leading role in narratives about human ancestors reflects the effect of the durability of stone on the availability of artifacts compared to plant matter. Recent analyses of wooden implements call attention to their importance as stand-alone tools and handles for mixed-material tools and testify to the mixture of hunting and nonhunting tools used in the distant past. Unsurprisingly, media reports take it for granted that all these tools were made by males, as in the *New York Times* headline, "The wood age? An analysis of 300,000-year-old wooden tools revealed that Neandertals were better craftsmen than previously thought."[5]

The headline refers to a 300,000 BP site near a lake in Schöningen, northern Germany, believed to have been inhabited by Neandertals or *H. heidelbergensis* (Leder et al. 2024). Different woods collected from faraway locations were used to create tools for gathering, hunting, and processing plant and animal matter. Of 187 wood specimens, there were 35 nonhunting tools for digging, scraping bark, and working and piercing hides. Another 17 tools of two types were made for hunting: 2-meter-long spears and lighter, smaller throwing sticks. Remains of horses as well as smaller animals and birds could indicate hunting in addition to scavenging, similarly to the processed elephant remains found at the same site as three yet-older wooden tools including a digging stick and a small finger-held implement from 430,000 BP that were found at Marathousa-1 in Greece (Milks et al. 2026). However, analysis of the bones of straight-tusked elephants found at Poggetti Vecchi in Central Italy indicates that the animals were butchered after they died from natural causes. Neandertal tools found with the bones and dated to 171,000 BP include boxwood implements over a meter long with pointed tips and rounded handles like the digging sticks used by recent and extant foraging populations throughout the world. The 50 remnants of these sticks indicate that their makers had charred them to render the tough wood easier to model with flint tools (Aranguren et al. 2018, 2019).

The discovery of wooden tools affirms that their makers had a mixed diet and used many nonhunting tools. In the case of the Gantangqing site in southwestern China, evidence of wooden tools and a subtropical rainforest environment suggest a predominantly vegetarian diet (Liu et al. 2025). Dated to 361,000 to 250,000 BP, the 35 wooden implements carved with stone blades include digging sticks, pointed hand tools, and hammers made of antlers. Starch grains on their tips indicate that the digging sticks were used to collect not only a wide variety of nuts, seeds, leaves, and fruits, but also sub-surface underground storage units (USOs) from the shores of an ancient lake.

The likelihood that both females and males manufactured hunting and nonhunting tools challenges the conventional assumption that toolmaking spaces in archeological sites were male zones whereas food processing and preparation spaces were occupied by females. Paradoxically, males are credited with having controlled fire, despite their presumed exclusion from cooking prehistorically and the contemporary pattern whereby women widely collect and cook carbohydrate-rich USOs including tubers, rhizomes, and bulbs. While the earliest evidence of cooking dates to a million years ago, when plant ashes and charred bone were left in Wonderwerk Cave, South Africa (Berna et al. 2012), it does not appear to have been widespread until around 200,000 BP. USOs were cooked at least 170,000 BP, as evidenced by charred remains of plant rhizomes found in Border Cave, South Africa (Wadley et al. 2020). A decline in the size of Neandertal and human faces and jaws reflects the impact of faster and easier mastication due to cooked food (Brace 1995). Cooking renders the carbohydrate in plants more easily absorbed and breaks down the collagen network in meat, increasing the efficiency of digestion. The energy spared by cooking food contributed to increased encephalization in humans and consequently played a role in cognitive modifications related to cooperative behavior. Fires also kept predators away and facilitated social interaction and pair bonding by extending the time for social interaction. The material presented in this chapter suggests that these important changes are likely to have been the handiwork of both females and males rather than one or the other.

## Plant and animal foods

As noted in Chapter 3, species in the *Paranthropus* and *Australopithecus* genera were herbivorous (although the latter may have eaten small sources of animal matter), like nonhuman primates. Species of *Homo*, in contrast, were omnivorous. Over several million years, the proportion of fat- and protein-rich animal parts in the diet increased dramatically, as did the need for more energy per day – much of it for fueling an outsized brain. The gastrointestinal tract shortened to less than two-thirds its expected length for body size compared to nonhuman primates, in line with the inverse relationship between gastrointestinal length and brain size across animal species. This length is shorter than it would be if humans were herbivores, but long relative to felines and other carnivores.

The shift to a diet consisting of both plant and animal foods left humans with a mix of traits shared with carnivores and herbivores (Lindeberg 2009). Like herbivores, humans have small mouths housing teeth with broad grinding surfaces, whereas carnivores have sharp teeth and large mouths and can swallow food whole. Carnivores synthesize vitamin C internally, whereas humans and herbivores must obtain the vitamin from dietary sources. Contrariwise, humans have insufficient enzymes that are produced internally by herbivores for synthesizing molecules such as the amino acid, taurine, or 20-carbon and 22-carbon polyunsaturated fatty acids (PUFAs); these molecules must be obtained through dietary sources and are abundant in meat. Humans also have genetic traits related to lipid metabolism and immune function that protect against cardiovascular and other diseases despite relatively high amounts of animal protein and fat in the diet (Finch and Stanford 2004).

The dietary intakes of recent and contemporary foragers reveal a consistent pattern of mixed food sources and a limit on protein consumption (see Eaton et al. 2002; Cordain et al. 2000). The liver is unable to clear the nitrogen in animal protein if consumption exceeds one-third of the calories in the diet, resulting in the accumulation of urea that harms kidney function and leads to liver enlargement and bone demineralization. Continued ingestion of excessive amounts of lean meat can be fatal within a few weeks. Among inland foraging populations, plant foods account for 50–80% of the food by weight. Staples include starchy USOs, protein-rich nuts and seeds, and edible leaves and stems; berries and fruit are relatively scarce and available only during short seasons. The rest of the diet, 20–50%, comes from animal sources, including insects, grubs, reptiles, fish, eggs, birds, and small and large game. Meat consumption is lowest among desert populations, at 15–20%. In coastal communities, up to 50% of the food comes from marine sources, the rest from plants. Arctic populations are exceptional in terms of animal sources of food, which reach 80–90% or more. Such high amounts are possible for several reasons, including consumption of organ meats, bone marrow, and large fat stores in marine mammals; the abundance of omega-3 PUFAs in foods obtained from the sea; and genetic variants mentioned in Chapter 3.

In arctic zones, men often supply nearly all the food through hunting. Exceptions include historical populations of Ainu foragers in northern Japan, among whom women's contribution of plant foods and trapped hares was at least as great as the meat provided by men. Where women's involvement in food procurement is limited, they nonetheless participate materially in other ways. For instance, those living in the far north of the American continent carry out the physically demanding labor of processing animals and hauling kill loads weighing more than 100 kilograms (Ocobock and Lacy 2024).

The ethnographic literature indicates that hunting is a routine activity for women across a wide range of foraging societies, lending strong support to the idea that females hunted throughout the course of human evolution. Anderson et al. (2023) drew a sample of nearly 400 out of 1,400 societies in the Database of Places, Languages, Culture, and Environment; consulted the original sources; and identified 63 societies in the Americas, Africa, Asia, Oceania, and Australia, for which there was explicit information about hunting.[6] For four-fifths of these societies (79%), the descriptions mention women actively hunting, as opposed to accompanying hunters. Of these, women's hunting is described as intentional, rather than opportunistic, in 87% of cases.

The descriptions indicate a variety of women's hunting styles, equipment, company, and size of game pursued. Opportunistic hunting involves only small game, whereas purposeful hunters tend overwhelmingly (with only two exceptions) to concentrate on small, medium, or large game – the last being the most common. Women bring children in half and dogs in 13% of the cases of purposeful hunting. They use many kinds of implements – including machetes, spears, knives, small and large sticks, crossbows, and nets – and they hunt alone or with other adults such as husbands or other women. Women hunt in 100% of the societies which consider hunting the most important subsistence activity.

Historically, Ainu women in northeast Asia brought dogs on large-game hunts. Women of indigenous populations of North America, such as the Ojibwa, Cree, and Chipewyan, hunted, fished, and trapped game. In Australia, Tiwi women hunted small game and made their own stone axes until Europeans brought metal implements. More recent and current examples include women Agta hunters of the Philippines, who hunt while menstruating, pregnant, and soon after childbirth, with the nursing infant in a sling. Women in the Nanadukan area spend as much time as men on hunting. They bring multiple dogs and carry bows and arrows. Hunting has no effect on childhood mortality, and having children has no effect on a woman's success at hunting. Women in the nearby Cagayan region go on extended hunting trips with their husbands, and both gather plant foods along the way.

In Africa, Ju/'hoansi women hunt by tracking; Aka and Mbuti women and men hunt with nets, with women in the former group capturing game while those in the latter are more likely to flush it out. In Papua New Guinea, Au and Gnau women kill animals they come across while gathering, whereas men hunt purposefully. In the Americas, Peruvian Matsés women and their husbands hunt together and bring in yields that exceed those of solo male hunters; among Aché foragers, hunting by tracking is the most time-consuming task for both women and men (Hill 2002).

An analysis of archeological assemblages together with samples of ethnographic foraging societies shows that recent gender divisions of subsistence and related tasks and technologies have replaced the Pleistocene pattern of nongendered food procurement (Zeanah et al. 2026).[7]

The Pleistocene-Holocene Transition (PHT) in the North American Great Basin provides an example of the kind of ecological and climatic shifts that set in motion a process leading to gender stratification. In the Pleistocene, humans in the area had access to plentiful high-quality edible plants and terrestrial and aquatic animal food in the area's vast wetlands. Females and males drove small game, fish, and fowl, and collectively hunted deer and large animals attracted to waterways and wetlands. The PHT brought a decline in the abundance of large game, which had to be procured by moving to new areas and engaging in more dangerous encounter hunting. Foragers continued to trap, fish, and eat seeds and USOs but mobility and the continued availability of high-quality foods kept them from investing in plant storage and processing. Thereafter, as wetlands shrunk due to the Holocene warming trend, the supply of both plants and game diminished, bringing shifts in technology to maximize the returns from collecting lower-yield plants and smaller animals, such as milling tools and hunting nets.

Based on the ethnographic datasets, the authors conclude that during the PHT males continued large-game hunting while females supported their efforts by transporting loads, constructing shelters, and processing game and hides, while concentrating on more reliable plant resources, fishing, trapping, and fowling. Thereafter, variegated gender divisions of labor arose in diverse locations as occurred throughout the world, in relation to each group's dependence on gathering, fishing, and hunting, as well as the proportions of the diet supplied by females and males. Analysis of the ethnographic sample leads the authors to argue that men consistently hunt large game regardless of their chances of success and contribute plant foods and smaller animals to varying degrees, whereas women consistently gather plant foods while engaging in hunting, trapping, and fishing to varying degrees. The authors maintain that women's choices are driven by the need to reliably provide for and protect children, although they acknowledge that childcare was not included in the analysis due to lack of suitable information in the datasets and that ethnographic sources tend to understate women's involvement in hunting.

Evidently, hunting and gathering are not mutually exclusive or limited to one sex or the other. The line between the two is blurry given that many forms of animal protein are gathered rather than hunted, such as shellfish, tortoises, eggs, or trapped animals. Hunting by women is not anomalous or unusual, nor is it incompatible with caring for children. Their generally greater involvement with plant foods and smaller animal sources indicates that women provide the most reliable source of nourishment in most societies. The work is physically demanding, whatever the balance of subsistence activities. Gathering involves covering similar distances to hunting and requires substantial effort, such as digging up to a meter beneath hard-packed soil for roots and tubers. Women regularly carry small children while hauling heavy loads of food, firewood, and water that add up to, or exceed, their body weight. As seen below, their bodies seem especially well designed for handling these tasks.

## Endurance capabilities

It is not just ethnographic evidence that upends the traditional evolutionary narrative about females being too weak and overburdened by children to engage in hunting. Together with paleopathological data, research on physical traits suggests that females are both highly capable of endurance exercise and likely to have been hunters for as long as males. The following discussion reports on a pair of articles by Lacy and Ocobock (2024) and Ocobock and Lacy (2024), along with additional sources cited separately.

Until around 40,000 BP, female and male skeletons show no differences in the locations or types of lesions, indicating that females and males engaged in the same activities and ran the same risks. The "rodeo rider" trauma injuries to the arms, head, and upper body in Neandertal males which are attributed to scuffles while subduing ambushed prey animals are present in the same locations on female skeletons. Equivalent arthritic lesions speak to equal levels of engagement in repetitive physical tasks, as well.

During the Late Pleistocene and Early Holocene, some skeletal differences indicate greater male involvement in hunting with projectiles. For instance, tendon insertion lesions in the elbow are twice as numerous on the right than left arm among Late Pleistocene males, but equally frequent on the two arms among females. However, while suggestive of projectile hunting, other causes of these "thrower's elbow" lesions cannot be ruled out. In addition, females are known to have engaged in projectile hunting, as seen in the case of a female buried beside projectiles and tools for processing animals around 9,000 BP at Wilamaya Patjxa in the Peruvian Andes. An analysis of prehistoric big-game hunters in the Americas suggests that up to half of them were females: among the sex-identified individuals buried with big-game hunting equipment such as spear throwers and projectile points at 18 sites, there were 11 females and 16 males (Haas et al. 2020).

The difference in the cross-sectional size of the humerus between the left and right arms indicates more use and strength in one arm compared to the other. On average this asymmetry is greater among Late Pleistocene males than females. However, humeral asymmetry increases with body size, rendering at least part of the apparent sex difference a question of size. In addition, there is substantial variation across populations, with cases of similar degrees of humeral asymmetry between females and males. For example, historical populations of nomadic foragers on the Aleutian Islands show both high robusticity and minimal sexual dimorphism in humeral asymmetry, indicating similar involvement of females and males in unimanual hunting and other strenuous activity necessitated by life in a harsh environment (Temple et al. 2021, 2023).[8]

Historical and archeological data attest to the participation of females in warfare during recent millennia, when armed conflict became more common (see Chapter 7). Females were buried with weapons and warrior equipment in one-third of cases among nomadic Scythians 2,500 BP. The high-status burial of Viking warrior/leader 1,200 to 1,000 BP with multiple types of weapons, two horses, shields, and tokens used for tactical planning determined to be a female indicates that the number of female warrior burials likely has been understated due to the assumption that individuals buried with weapons were males (see Anderson et al. 2023).

Today's women soldiers and athletes equal men in shooting ability.[9] Women recruits perform as well as men on physical tests now that these have shifted from timed pushups and three-mile runs to skills appropriate to the work, such as walking long distances at night through bad weather while carrying a heavy load and then setting up an ambush; loading multiple mortar rounds onto a truck; or swimming oneself and a wounded colleague to safety while wearing a flight suit (McGrath and O'Hanton 2025). Women have been flying planes for the US Air Force since the 1970s, including fighter planes starting in the 1990; over the same half-century, death rates from accidents in the military have steadily declined, confirming the lack of a negative effect on military effectiveness (McGrath and O'Hanton 2025).[10]

As noted in the last chapter, the size difference between women and men is greater among the taller populations of temperate zones than equatorial or arctic latitudes. This suggests that ancestral females and males were closer in size than today's midlatitude populations, which are themselves much taller today than they were in the past. Being closer in size would mean that they were closer in strength, given that men's greater strength is due to more muscle mass rather than a sex difference in the power of muscle tissue. In addition, contemporary societies that limit or prohibit the participation of girls and women in sports and exercise create wider gender differences in musculature than ones in which everyone is physically active.

Humans of all sexes have anatomical and physiological traits that allow for extended periods of endurance exercise, such as persistence hunting (see Morin and Winterhalder 2024). Women seem to be especially well-suited for long-distance physical events, including running and bicycling races. They perform similarly to men in marathons and outpace them in running races longer than 64 kilometers. Women's wider pelvis relative to height is not a disadvantage given that women rotate their hips more than men while walking, which extends stride length and makes walking more efficient. The wider pelvis also allows women to carry loads more efficiently than men. In experimental conditions women move faster, with longer strides, while walking and carrying a simulated or real baby. They use up to a third less energy than men, even though the load is greater relative to their body weight.

Women's muscles contain a greater proportion of Type 1 (slow-twitch) fibers than men's. These fibers rely on anaerobic metabolism, which is a slower way to create energy as is needed for long periods of exercise. Thanks to Type 1 fibers, when women stretch and then shorten muscle – as in a squat followed by a pause and finally a jump – their muscles utilize 90% of the energy produced during the squat compared to men's 50%. A higher proportion of Type 1 muscle fibers also renders the muscles less prone to fatigue, as seen in differences between women and men's performance in weightlifting. Women can complete more repetitions than men at equivalent proportions of their personal maximal load (32 more repetitions at 50% of the maximum, 13 more at 60%, and 5 more at 70%).

In addition, women's muscles oxidize 70% more fatty acids than men's. Compared to carbohydrates, fatty acids are a more slowly utilized and denser source of energy (four versus nine calories per gram, respectively). Glycogen stores are kept in reserve for later use during endurance activity, delaying fatigue. The reason for women's greater use of fatty acids during exercise seems to be higher levels of a form of estrogen, 17-beta-estradiol or E2. Training through endurance exercise renders men's muscle cells more sensitive to E2 by increasing the number of E2 receptors. More E2 also gives women's muscles greater resistance to the breakdown of muscle protein, resulting in less urea excreted after exercise, and causes the amino acid leucine, which is needed for muscle growth, to decay at a lower rate. Men given E2 experience a reduction in leucine breakdown.

Despite carrying more subcutaneous body fat, women have lower rates of cardiometabolic disorders than men. Due to hormonal differences fat tends to be deposited on the hips and thighs in women but the abdomen in men, where increased amounts correlate with higher risks of cardiometabolic disorders. In addition, women's bodies produce up to 65% more of the hormone adiponectin, which regulates fatty acid and glucose metabolism and thereby protects against non-insulin-dependent diabetes. Greater fatty acid utilization largely explains women's being able to engage in endurance (as well as resistance) exercise for longer than men before experiencing fatigue. In addition, women keep a more constant speed over long-distance running races than men, although some of this may be due to psychological factors. They recover from exercise sooner than men and their cells are better protected against damage due to lower levels of heat-shock protein and creatine kinase from stress-damaged cells. This effect seems to be mediated by E2, which also protects cell membranes due to its antioxidant properties.

In consideration of these and other physical traits, Ocobock and Lacy (2024) conclude that humankind's ability to sustain endurance exercise may derive from adaptations related to ancestral females' success at providing for themselves during pregnancy and lactation while also dealing with potentially limited food sources and, in the distant past, predators. Women need an additional 500 calories per day during the last trimester of pregnancy and 600 extra calories per day during lactation. Characteristics that conferred advantages for endurance activity would have allowed ancestral females to meet this need. In fact, among both female and male athletes today, endurance training results in some of the same changes that occur during pregnancy, such as greater fatty acid utilization and increased blood volume with attendant cardiac modifications. That is, not only is reproduction compatible with physically demanding tasks but it could be reason all humans have traits favorable to endurance activity.

Table 6.1 summarizes the anti-essentialist implications of the material presented so far as well as prehistoric creative expression, discussed below.

*Table 6.1* Ethnographic and Archeological Evidence against Traditional Narratives about Prehistory

| | *Characteristics* | *Implications* |
|---|---|---|
| Demography | • low relatedness within groups<br>• bilateral kinship and bilocal residence | • groups not organized around related males<br>• cooperation not rooted in warfare/raiding |
| Cooperation | • childbirth assistance<br>• alloparenting<br>• care of disease and disability | • mutual assistance and protection as the basis of cooperation<br>• females and males both involved in child care |
| Toolmaking | • non-hunting and hunting tools<br>• wood and stone<br>• made for personal use | • toolmaking by all<br>• recent origin of hunting implements<br>• no basis for assumptions about male visuospatial abilities due to and manifested in toolmaking |
| Food production | • biological characteristics of omnivores<br>• forager diets<br>• female hunters | • variability in food sources<br>• importance of vegetable foods<br>• females not provisioned by males |
| Female endurance capabilities | • traits favorable to endurance activity while carrying weights<br>• female participation in warfare | • hunting and warfare not incompatible with reproduction<br>• human endurance capabilities may be rooted in adaptations related to pregnancy |
| Creative expression | • indications of female authorship<br>• monumental art | • likely female role in artistic expression and spiritual leadership<br>• against image of prehistoric life as brutal, violent, and limited to the struggle for survival |

**Creative expression**

The equation of males with visuospatial ability and toolmaking noted above is implicit in the assumption that prehistoric art forms were created by males. For instance, the so-called "Venus figurines," produced between 35,000 and 11,000 BP but mostly from 26,000 to 21,000 BP (Gravettian period/style), have traditionally been considered pornographic objects made by males for their own enjoyment. Hundreds of these ivory, soft stone, and clay statuettes have been found in multiple sites across Eurasia, a minority of which represent males. No pornographic purpose has been suggested for the latter. Given the dearth of evidence that past societies might have viewed female nudity in the sexualized way of contemporary pornography, archeologists search for alternative explanations. The figurines could have had a didactic purpose in teaching girls about reproduction, or pregnant women may have made them as self-portraits. The latter would explain the preponderance of statuettes with featureless faces and large breasts, bellies, and hips, but small feet (Lacy and Ocobock 2024).

That some figurines have been found in fire pits, intentionally burned to the point of cracking, suggests that they could have been used in divination, much like animal scapulae (Nowell and Chang 2014). The figurines could have expressed an ideal of female health and beauty during times of scarcity due to climate change, for plumper figurines were made by humans living nearer to advancing glaciers and during periods closer to the glacial maximum (Johnson et al. 2021). Textile impressions could indicate female manufacture given the widespread although not universal association of women with weaving (Soffer et al. 2000). In short, the figurines could have expressed many things and been made by females, males, or both.

The same could be said of the many other kinds of decorative objects made by human ancestors. Beginning around 80,000 BP, in Africa and the Middle East they produced decorative beads, pigments and paints, and etchings on ochre and ostrich eggs. Starting around 43,000 BP, sites in Europe indicate abundant creative expression and human interaction, such as Lascaux and Grotte Chauvet in France, El Castillo in Spain, and the Grotta del Fumane in Italy.[11] During the Upper Paleolithic, trade networks of shells, stones, and beads linked groups over great distances and brought them together periodically in huge numbers, as in the Dordogne River Valley for beads, or the Kalahari Desert for personalized projectile points; notation systems were developed, possibly to keep track of time and seasons; and humans created monuments, musical instruments including perforated bone flutes, and more specialized and sophisticated tools (see Bar-Yosef 2002; d'Errico and Banks 2013).

Upper Paleolithic artists engaged in visual storytelling and depicted recognizable animals on rock walls. Despite their evident skill, they refrained from painting equally naturalistic humans, possibly reflecting beliefs about the spiritual danger of representing a human face shared by some cultural and religious traditions today. That the artists painted in inaccessible caves, tight spaces, deep reaches, and high ledges further suggests a religious/spiritual motive; that they painted and sculpted hybrid animal-animal and animal-human forms, such as the 40,000-year-old "Lion Man" standing figurine from Stadel Cave in southwest Germany (human lower body; head and arms of a lion), suggests they had the imagination to conjure up creatures that did not exist and consequently the cognitive capacities at the root of religious belief.

On the Indonesian island of Sulawesi, humans created parietal art even before the arrival of *H. sapiens* in Europe. Paintings that date to 48,000 BP or earlier at one site (Leang Bulu' Sipong 4) and 51,200 BP at another (Leang Karampuang) (Oktaviana et al. 2024) show tiny stick-figure humans surrounding native species of warty pigs and horned dwarf buffalo drawn exceedingly large by comparison. Some of the human figures carry a spear or rope, and some of their heads have nonhuman features such as a snout, beak, or tail. They could be hybrid creatures or hunters wearing a prosaic disguise, a reminder of the caution needed in interpreting archeological materials.

Despite the uncertainties involved, the artistic and technological advances listed above are routinely attributed to males and assumed to be about males and male-associated topics (see Chapter 4). Analysis of the art tells a different story (Lacy and Ocobock 2024). For instance, no images have been found to represent human-on-human violence, which is not surprising given the lack of evidence for Paleolithic warfare. Humans are rarely represented; they tend to be drawn as stick figures; and few are engaged in an obvious and unmistakable task. Some images show injured animals with no human figures present; others show humans who have been injured by an animal. Where cave paintings clearly concern hunting – by including both animals and humans with implements – the figures are generic humans, as in the paintings described above. Cave paintings of a more complete human figure appear beginning 16,000 BP, yet many do not clearly suggest its sex. For instance, the red-pigment rock painting of a dog at the heels of a human in Libya, dated to 7,000 to 4,000 BP, shows a person walking with arms outstretched, a stick in one hand, something oval-shaped dangling from the other (in Grimm 2020). The figure has a diminutive head, wide chest, very narrow and long waist, and wide skirt-covered hips above muscular legs.

During the Late Pleistocene and Early Holocene, foragers made monumental structures and art forms that testify to the existence of political/administrative organization unrelated to agriculture and permanent settlement, discussed in the next chapter. The five-meter-long Shigir Idol, made of a single split larch plank near Yekaterinburg, Siberia, dates to 10,200 BP, a time when glaciers were receding from Eurasia and woodlands were expanding (Terberger et al. 2021). Eight faces adorn its front and back, including a three-dimensional head with eyes, a nose, and a mouth pursed into an "o." In a departure from Paleolithic representational art, there are geometrical patterns of simple lines, chevrons, and zigzags. Starting around 11,000 BP, at Göbekli Tepe in Turkey, foragers built and modified 20 enormous stone enclosures that include carved T-shaped limestone pillars of a similar height of five or more meters. Dangerous reptiles, carnivores, and scavengers – many with prominent male sexual organs – adorn the pillars, some protruding in three dimensions. There are many raptors and other creatures with human heads in their grasp. Other pillars stand like humans, with limbs and clothing. The pillars may have held up roofed buildings, where people congregated in summer and fall to take advantage of animal migration routes and plant maturation times. Like paintings of hybrid creatures, the awe-inspiring images suggest an otherworldly purpose, such as protection from dangerous spirits.

Along a zone beneath the glaciers, from Eastern Europe to Russia, foragers constructed large circular monuments out of mammoth bones and tusks starting around 25,000 BP. The amount of animal material they collected suggests that hundreds of people gathered there to hunt, celebrate, and trade or gift goods (Soffer 2013). In the Americas, the number of people and size of monuments was far greater. For instance, at Louisiana's Poverty Point on the lower Mississippi, from 3,700 to 3,100 BP foragers left caches of colorful stones, tools, vessels, arrowheads, and clay balls presumably used in cooking. All the mineral and lithic elements came from elsewhere, including places hundreds of miles away. The meaning of the ridges, plaza, and mounds remains unclear, but their scale and geometrical designs clearly suggest the work of thousands of people, while the similarity of the geometrical layout of the vast settlement to others across the region points to the likelihood that this was a place for feasting and ritual or ceremonial events (Hays 2019).

While it is impossible to know the sex of the people who made the art discussed above, one type of painting provides clues: handprints made with tints or by blowing paint around a hand against a rock wall. Stencils and handprints from 37,000 years ago are common in parietal art throughout Europe (for instance, Rouffignac, Chauvet, Gargas, and El Castillo, to name a few). The ones on Sulawesi are 40,000 years old. An analysis of hand impressions in 59 caves in Europe and the UK indicates that they were made by females and males, and that the artists included infants, children, adults, and older adults (Fernàndez-Navarro et al. 2025). While the study estimated that more of the prints were made by males than females, an analysis of handprints in El Castillo found balanced numbers (Rabazo-Rodríguez et al. 2017), and a study of caves in France and Spain determined that the prints were majority female (Snow 2013). These studies indicate that the size of the fingers and palms is more predictive than either the ratio between the second and fourth fingers (lower in males due to prenatal androgen exposure in fetal life) or the shape of the hand as characterized through dozens of measurements. Although exactitude in estimating proportions is unattainable due to uncertainties stemming from differences between Paleolithic and modern hands, the types of measurements used, and the applicability of methods across settings (see also Galeta et al. 2014), it is nonetheless clear that parietal art was not made entirely or even mostly by males.

From the foregoing it seems safe to conclude that Upper Paleolithic art was made by both females and males. Considering the evidence of female endurance capabilities and women's hunting among extant foraging societies, there is no reason to imagine that hunting, or art about hunting, concerned only males. Hybrid forms and monumental art suggest ritual functions and the existence of spiritual leaders, who, as seen below, could well have been female.

## Holocene foragers

As group size increased in the Upper Pleistocene and Holocene, specialization became more practicable for foraging societies. As seen in the next chapter, this did not automatically cause stratification and inequality.

A couple of burials attest to continued high female status and lend further support to the principle that female foragers engaged in the same tasks as males, including hunting. The first, from 5,000 BP in Valencia, Spain, was initially classified as male based on pelvic measurements, which dovetailed with gender expectations sparked by the enormous quantity of precious grave goods including a 1.8-kilogram African elephant tusk; several ceramic plates, one of which is tinged with cannabis and wine; an ivory vessel; flint blades; and, a rock-crystal dagger whose handle is decorated with 90 perforated mother-of-pearl beads. A re-examination of the skeletal remains detected the presence of the tooth enamel protein, amelogenin, that is found only in females (Cintas-Peña et al. 2023). "Ivory Lady" died at age 17–25 years. Mercury in her bones indicates that she was exposed to the red pigment, cinnabar, during her lifetime. Some of the grave goods were added two to three generations after the initial interment, and a set of carefully placed flat slabs was placed over it. Nearby, around 100 meters from the grave, the same people who took care of the grave – possibly the Ivory Lady's descendants – built a two-chambered tomb filled with lavish grave goods over the period from around 4,900 to 4,600 BP. The small chamber contains the remains of 5 people and is badly damaged; the large one contains the remains of 15 female or likely female individuals and 5 individuals of undetermined sex, all 20–35 years of age. Some are clothed in garments decorated with thousands of seashell beads. Their bones contain high levels of mercury from antemortem exposure to cinnabar. The inclusion of one female with six toes on each foot further suggests these females' religious role, as this type of anomaly would have given her special status.

The Ivory Lady seems to have been the most important personage in the Iberian Peninsula for several hundred years. Her standing was likely to have been achieved rather than ascribed, is typical of small-scale societies (see Chapter 7), and evident from the lack of infant burials in the area that contained grave goods. That her position was unmatched by anyone else's, male or female, is suggested by the complete absence of equally lavish graves. Other females were buried with valuable grave goods during the Copper and Iron Ages in Europe (such as Franzhausen and La Almoloya), yet no males in their communities received comparable treatment. Evidently, the most highly esteemed person in the region could be a female.

The second example concerns Burial XXII or "Sitting Woman," a female aged 30–40 years interred around 7,000 BP in Skateholm, southern Sweden, atop a bed of deer antlers. A cloak of feathers from birds including jays, seagulls, ravens, and ducks and geese had been placed around her shoulders. From her neck hung an imposing pendant made of slate, and around her waist there was belt composed of over 100 elk, deer, and boar teeth. These objects indicate that Sitting Woman was an important person, possibly a healer and/or spiritual leader. Originally excavated by Lars Larsson among multiple graves in the 1980s, her DNA was analyzed in 2019 and showed that she was genetically similar to the foragers buried nearby. From this information Oscar Nilsson (2026) created a life-sized reconstruction that depicts a woman with pale eyes and dark skin (see Chapter 3 regarding the use of aDNA in reconstructing the past).

Sitting Woman's group lived in a resource-rich region with lowlands and forests where they could have practiced agriculture, and they would have had the knowledge thanks to contact with agriculturalists across the water in Europe, with whom they traded. Instead, they foraged, fished, and hunted, and buried their dogs with grave goods or alongside people. Like Ivory Lady, Sitting Woman indicates that females could be revered and hold positions of great prestige even as many populations were trending toward greater social and gender inequality. Egalitarianism and the persistence of foraging on the part of some populations despite knowledge of agriculture are discussed further in the next chapter.

## Gender and human evolution

Contrary to essentialist beliefs about evolutionary processes centered on hunting and toolmaking by males as the basis of gender differences, this chapter has shown that females are likely to have obtained animal meat for as long as males, and the female body is up to the task. Until late in the Pleistocene, groups of human ancestors were numerically too small for specialization, requiring everyone to be a generalist. Once group size increased, males seem to have done more projectile hunting on average, yet later inventions such as the spear thrower, bow, fishhooks, and hunting and fishing nets could be used effectively by both females and males. Everyone needed non-hunting tools, which were more common than tools for hunting. The continued practice of hunting by females in recent and extant foraging societies suggests that use of any kind of tool would not have been restricted to males.

The last chapter demonstrated that human females and males do not have separate, non-overlapping cognitive abilities. Together these two chapters remove the foundation beneath the idea that evolutionary processes resulted in genetically determined sex-specific cognitive traits rooted in divergent food-procurement strategies. Indeed, the entire focus on subsistence seems misplaced considering the variability in food sources, the integration of food collection with other activities of daily life, and the fact that foraging occupies less time and requires less physical effort than agriculture (see Cohen 1989; Dahlberg 1981).

The emergence of food production and large, permanent settlements brought more specialization relative to foraging societies, allowed some individuals and groups to gain power over others, and increased the reasons for conflict within and between groups. However, foraging societies did not turn permanently and definitively to agriculture all in one go, nor did agriculture uniformly bring social and gender stratification, political consolidation, and wealth concentration. Many early agricultural societies combined foraging with food production and remained relatively egalitarian. Large urban settlements without elites or military installations persisted for hundreds to thousands of years. Some populations developed political centralization and great wealth inequality only to later abandon them for more egalitarian political and socioeconomic systems.

Today, most of the world's population lives in stratified societies with cultural beliefs to match. Its seeming basis in biological differences makes genetic essentialism a highly effective instrument in the service of inequality even though, as shown in the next chapter, the social inequality it upholds is anything but natural or inevitable.

## Notes

1 Age reduces this width and renders the pelvic bones thinner and more brittle, while childbearing increases it.
2 These differences are not due to early death among foragers. People in foraging and horticultural societies around the world today live to the age of 68–78,

which may represent an approximate life span for modern *H. sapiens* (Gurven and Kaplan 2007). Estimations based on comparisons with other primate species yield a longer human life span of 90–95 years (Cutler 1975).

3 Instead of the "obstetrical dilemma" between upright posture and encephalization, maternal metabolic constraints could explain the timing of birth in humans, as they do in other species (Dunsworth 2016; Dunsworth et al. 2012). The wider female pelvis therefore represents a compromise between this energetic calculus and the cost to locomotive efficiency should it have expanded yet further. Support for the metabolic hypothesis comes also from the fact that, compared to closely related primate species, human gestation runs longer than expected, not shorter. In addition, human babies are the largest in absolute body and brain size.

4 Regarding expected age at weaning for humans, see Dettwyler (1995).

5 6 May 2024.

6 See also Lacy and Ocobock (2024).

7 *n* = 43 societies from the Standard Cross-Cultural Sample; *n* = 103 societies from the Western North American Indians database.

8 Other populations living during the same time span (1,400–200 BP) in Alaska present a variegated picture of strength and humeral asymmetry related to different tasks typical of northern areas, inland zones, and a coastal fishing village.

9 For an analysis of target shooting among Olympic athletes, see Goldschmied et al. (2023).

10 To illustrate, there were 72 deaths per 100,000 in 1980; 28 in 2000; and 21.5 in 2020 (Defense Casualty Analysis System 2025).

11 Neandertals also created art forms, such as jewelry and storage containers out of shells 50,000–45,000 BP in Spain (Zilhão et al. 2010). They may have painted the waterfall panel in Pasiega cave, Spain, which has been dated to 65,000 BP, although White et al. (2020) question the method of comparing the ratio of uranium to its thorium decay product in the calcite left by dripping groundwater. The leaching of uranium over time, among other factors, may cause overestimation of the age of parietal art.

## References


Adovasio JM, Soffer O, Page J. 2016. *The invisible sex*. Routledge.

Amoss PT, Harrell S. 1981. *Other ways of growing old*. Stanford University Press.

Anderson A, Chilczuk S, Nelson K, Ruther R, Wall-Scheffler C. 2023. The myth of Man the Hunter. *PLoS One* 18(6):e0287101.

Aranguren B, Revedin A, Amico N, Cavulli F, Giachi G, et al. 2018. Wooden tools and fire technology in the early Neanderthal site of Poggetti Vecchi (Italy). *Proceedings of the National Academy of Sciences* 115(9):2054–2059.

Aranguren B, Grimaldi S, Benvenuti M, Capalbo C, Cavanna F, et al. 2019. Poggetti Vecchi (Tuscany, Italy). *Journal of Human Evolution* 133:32–60.

Arthur KW. 2010. Feminine knowledge and skill reconsidered. *American Anthropologist* 112(2):228–243.

Bar-Yosef O. 2002. The Upper Paleolithic revolution. *Annual Review of Anthropology* 31:363–393.

Beckerman S, Valentine P, editors. 2002. *Cultures of multiple fathers*. University Press of Florida.

Ben-Dor M, Barkai R. 2023. The evolution of Paleolithic hunting weapons. *Quaternary* 6(3):46.
Bergström A, Frantz L, Schmidt R, Ersmark E, Lebrasseur O, et al. 2020. Origins and genetic legacy of prehistoric dogs. *Science* 370(6516):557–564.
Berna, F., Goldberg, P., Horwitz, L. K., Brink, J., Holt, S., et al. 2012. Microstratigraphic evidence of in situ fire in the Acheulean strata of Wonderwerk Cave, Northern Cape province, South Africa. *Proceedings of the National Academy of Sciences 109*(20), E1215–E1220.
Brace CL. 1995 [1967]. *The stages of human evolution.* 5th ed. Prentice-Hall.
Cintas-Peña M, Luciañez-Triviño M, Montero Artús R, Bileck A, Bortel P, et al. 2023. Amelogenin peptide analyses reveal female leadership in Copper Age Iberia. *Scientific Reports* 13:9594. https://doi.org/10.1038/s41598-023-36368-x
Cohen MN. 1989. *Health and the rise of civilization.* Yale University Press.
Cohn BA. 1998. The vital role of the skin in human natural history. *International Journal of Dermatology* 37(11):821–824.
Cordain L, Miller JB, Eaton SB, Mann N, Holt SH, et al. 2000. Plant-animal subsistence ratios and macronutrient energy estimations in worldwide hunter-gatherer diets. *American Journal of Clinical Nutrition* 71(3):682–692.
Cutler RG. 1975. Evolution of human longevity and the genetic complexity governing aging rate. *Proceedings of the National Academy of Sciences* 72(11):4664–4668.
d'Errico F, Banks WE. 2013. Identifying mechanisms behind Middle Paleolithic and Middle Stone Age cultural trajectories. *Current Anthropology* 54(S8):S371–87.
Dahlberg F, editor. 1981. *Woman the gatherer.* Yale University Press.
Defense Casualty Analysis System. 2025. U.S. active military duty death rates per 100,000 serving. https://dcas.dmdc.osd.mil/dcas/app/summaryData/deaths/numberServe
Dettwyler K. 1995. A time to wean. In Stuart-Macadam P, Dettwyler KA, editors. *Breastfeeding.* Aldine de Gruyter. pp. 39–73.
Dunsworth HM. 2016. Thank your intelligent mother for your big brain. *Proceedings of the National Academy of Sciences* 113(25):6816–6818.
Dunsworth HM, Warrener AG, Deacon T, Ellison PT, Pontzer H. 2012. Metabolic hypothesis for human altriciality. *Proceedings of the National Academy of Sciences* 109(38):15212–15216.
Dyble M, Salali GD, Chaudhary N, Page A, Smith D, et al. 2015. Sex equality can explain the unique social structure of hunter-gatherer bands. *Science* 348(6236):796–798.
Eaton SB, Pike MC, Short RV, Lee NC, Trussell J, et al. 1994. Women's reproductive cancers in evolutionary context. *The Quarterly Review of Biology* 69(3):353–367.
Eaton SB, Strassman BI, Nesse RM, Neel JV, Ewald PW, et al. 2002. Evolutionary health promotion. *Preventive Medicine* 34(2):109–118.
Fernàndez-Navarro VF, Casares DF, Martinez DG, Maidagan DG. 2025. Decoding Palaeolithic hand stencils. *Journal of Archaeological Method and Theory* 32(1):24.
Finch CE, Stanford CB. 2004. Meat-adaptive genes and the evolution of slower aging in humans. *The Quarterly Review of Biology* 79(1):3–50.
Fry DP. 2006. *The human potential for peace.* Oxford University Press.
Galeta P, Bruzek J, Lázničková-Galetová M. 2014. Is sex estimation from handprints in prehistoric cave art reliable? *Journal of Archaeological Science* 45:141–149.

Geller PL. 2009. Bodyscapes, biology, and heteronormativity. *American Anthropologist* 111(4):504–516.
Gettler LT. 2010. Direct male care and hominin evolution. *American Anthropologist* 112(1):7–12.
Gettler LT, McKenna JJ, McDade TW, Agustin SS, Kuzawa CW. 2012. Does cosleeping contribute to lower testosterone levels in fathers? *PLoS One* 7(9):e41559.
Goldschmied N, García-Aliaga A, Muriarte Solana D, Mon-López D. 2023. Do women and men compete equally on a level playing field? *PLoS One* 18(9):e0291017.
Grimm D. 2020. How dogs tracked their humans across the ancient world. *Science* 370(6516):522.
Gurven M, Kaplan H. 2007. Longevity among hunter-gatherers. *Population and Development Review* 33(2):321–365.
Haas R, Watson J, Buonasera T, Southon J, Chen JC, et al. 2020. Female hunters of the early Americas. *Science Advances* 6(45):eabd0310.
Hart D, Sussman RW. 2005. *Man the hunted*. Basic Books.
Hawkes K, O'Connell JF, Jones NB, Alvarez H, Charnov EL. 1998. Grandmothering, menopause, and the evolution of human life histories. *Proceedings of the National Academy of Sciences* 95(3):1336–1339.
Hays CT. 2019. Feasting at Poverty Point with Poverty Point objects. *Southeastern Archaeology* 38(3):193–207.
Hewlett BS. 2017. *Father-child relations*. Routledge.
Hill K. 2002. Altruistic cooperation during foraging by the Ache, and the evolved human predisposition to cooperate. *Human Nature* 13:105–128.
Hill KR, Walker RS, Božičević M, Eder J, Headland T, et al. 2011. Co-residence patterns in hunter-gatherer societies show unique human social structure. *Science* 331(6022):1286–1289.
Hill KR, Wood BM, Baggio J, Hurtado AM, Boyd RT. 2014. Hunter-gatherer interband interaction rates. *PloS one*. 9(7):e102806.
Hrdy SB. 2011. *Mothers and others*. Harvard University Press.
Jaeggi AV, Gurven M. 2013. Natural cooperators. *Evolutionary Anthropology* 22(4):186–195.
Johnson RJ, Lanaspa MA, Fox JW. 2021. Upper Paleolithic figurines showing women with obesity may represent survival symbols of climatic change. *Obesity* 29(1):11–15.
Lacy S, Ocobock C. 2024. Woman the hunter. *American Anthropologist* 126(1):19–31.
Leder D, Lehmann J, Milks A, Koddenberg T, Sietz M, et al. 2024. The wooden artifacts from Schöningen's Spear Horizon and their place in human evolution. *Proceedings of the National Academy of Sciences* 121(15):e2320484121.
Lee RB, DeVore I, editors. 1968. *Man the hunter*. Routledge.
Lindeberg S. 2009. Modern human physiology with respect to evolutionary adaptations that relate to diet in the past. In Hublin JJ, Richards MP, editors. *The evolution of hominin diets*. Springer. pp. 43–58.
Liu JH, Ruan QJ, Ge JY, Huang YJ, Zhang XL, et al. 2025. 300,000-year-old wooden tools from Gantangqing, southwest China. *Science* 389(6755):78–83.
Marsh WA, Scarsbrook L, Yüncü E, Hodgson L, Lin AT, et al. 2026. Dogs were widely distributed across western Eurasia during the Palaeolithic. *Nature* 651(8107):995–1003.
Martin M, de la Mora AN, Valeggia C, Veile A. 2024. Can women hunt? *American Anthropologist* 126(2):365–369.

McGrath A., O'Hanton M. 2025. Think the military lowered its standards for women? *Washington Post* 11 February 2025.
Milks A, Ntinou M, Koutalis S, Michailidis D, Konidaris GE, et al. 2026. Evidence for the earliest hominin use of wooden handheld tools found at Marathousa 1 (Greece). *Proceedings of the National Academy of Sciences* 123(6):e2515479123.
Morin E, Winterhalder B. 2024. Ethnography and ethnohistory support the efficiency of hunting through endurance running in humans. *Nature Human Behaviour* 8:1065–1075.
Nilsson OD. 2026. A 7,000-year-old woman from Skateholm, Sweden. https://www.odnilsson.com/gallery-skateholm/
Nowell A, Chang ML. 2014. Science, the media, and interpretations of Upper Paleolithic figurines. *American Anthropologist* 116(3):562–577.
Ocobock C, Lacy S. 2024. Woman the hunter. *American Anthropologist* 126(1):7–18.
Oktaviana AA, Joannes-Boyau R, Hakim B, Burhan B, Sardi R, et al. 2024. Narrative cave art in Indonesia by 51,200 years ago. *Nature* 631(8022):814–818.
Pilli E, Vai S, Moses VC, Morelli S, Lari M, et al. 2024. Ancient DNA challenges prevailing interpretations of the Pompeii plaster casts. *Current Biology*. https://doi.org/10.1016/j.cub.2024.10.007
Rabazo-Rodríguez AM, Modesto-Mata M, Bermejo L, García-Díez M. 2017. New data on the sexual dimorphism of the hand stencils in El Castillo Cave (Cantabria, Spain). *Journal of Archaeological Science* 14:374–381.
Rilling JK, Young LJ. 2014. The biology of mammalian parenting and its effect on offspring social development. *Science* 345(6198):771–776.
Rilling JK, Lee M, Zhou C, Jung E, Arrant E, et al. 2025. Hormonal changes in first-time human fathers in relation to paternal investment. *Hormones and Behavior* 171:105740.
Schrenk A, Tilley L. 2018. *Caring in ancient times. Anthropology News* 59(1):e57–63. https://doi.org/10.1111/AN.743
Sear R, Mace R. 2008. Who keeps children alive? *Evolution and Human Behavior* 29(1):1–18.
Shepard GH Jr, Levi T, Neves EG, Peres CA, Yu DW. 2012. Hunting in ancient and modern Amazonia. *American Anthropologist* 114(4):652–667.
Sikora M, Seguin-Orlando A, Sousa VC, Albrechtsen A, Korneliussen T, et al. 2017. Ancient genomes show social and reproductive behavior of early Upper Paleolithic foragers. *Science* 358(6363):659–662.
Snow DR. 2013. Sexual dimorphism in European Upper Paleolithic cave art. *American Antiquity* 78(4):746–761.
Soffer O. 2013. *The Upper Paleolithic of the central Russian plain*. Elsevier.
Soffer O, Adovasio JM, Hyland DC. 2000. The "Venus" figurines. *Current Anthropology* 41(4):511–537.
Sterling K. 2015. Man the hunter, woman the gatherer? In Cummings V, Jordan P, Zvelebil M, editors. *The Oxford handbook of the archaeology and anthropology of hunter-gatherers*. Oxford University Press. pp. 151–173.
Temple DH, Rosa ER, Hunt DR, Ruff CB. 2021. Adapting in the Arctic. *American Journal of Physical Anthropology* 176(1):3–20.
Temple DH, Rosa E, Hunt DR, Ruff CB. 2023. Adapting in the Arctic II. *American Journal of Biological Anthropology* 181(3):392–412.
Terberger T, Zhilin M, Savchenko S. 2021. The Shigir idol in the context of early art in Eurasia. *Quaternary International* 573:14–29.

Townsend C, Aktipis A, Cronk L. 2021. Does scarcity lead to selfishness? *Anthropology News* 62(3):8–12.
Trevathan W. 2015. Primate pelvic anatomy and implications for birth. *Philosophical Transactions of the Royal Society B: Biological Sciences* 370(1663):20140065.
Tronick EZ, Morelli GA, Winn S. 1987. Multiple caretaking of Efe (Pygmy) infants. *American Anthropologist* 89(1):96–106.
Vinicius L, Mace R, Migliano A. 2014. Variation in male reproductive longevity across traditional societies. *PLoS One* 9(11):e112236.
Wadley L, Backwell L, d'Errico F, Sievers C. 2020. Cooked starchy rhizomes in Africa 170 thousand years ago. *Science* 367(6473):87–91.
Walker RS, Flinn MV, Hill KR. 2010. Evolutionary history of partible paternity in lowland South America. *Proceedings of the National Academy of Sciences* 107(45):19195–19200.
White R, Bosinski G, Bourrillon R, Clottes J, Conkey MW, et al. 2020. Still no archaeological evidence that Neanderthals created Iberian cave art. *Journal of Human Evolution* 144:102640.
Wiessner P. 1996. Leveling the hunter. In Wiessner P, Schiefenhövel W, editors. *Food and the status quest*. Berghahn Books. pp. 171–191.
Zeanah DW, Elston RG, Codding BF. 2026. Gender convergent labor and technological change at the Pleistocene to Holocene Transition in the Great Basin. *Journal of Anthropological Archaeology* 81:101746.
Zilhão J, Angelucci DE, Badal-García E, d'Errico F, Daniel F, Dayet L, et al. 2010. Symbolic use of marine shells and mineral pigments by Iberian Neandertals. *Proceedings of the National Academy of Sciences* 107(3):1023–1028.

# 7 How did we get here?

## Inequality and essentialism

It would be easy to gather from news reports that humans are prone to hierarchical, other-avoiding, and aggressive behavior. This vision of humankind's default setting reverberates through multiple academic fields: psychological experiments verify that small children can recognize and enact inequality in interpersonal interactions; primate dominance hierarchies and especially those of chimpanzees are taken to model evolutionarily driven impulses and natural male dominance; intertribal violence among ethnographic societies is offered as evidence that groups of humans were territorial and warlike in prehistory as well. The "great men and events" historical tradition keeps the spotlight on wars, geopolitical divisions, and the brutal collapse of civilizations. These dark interpretations of human biology and history resonate in an extremely unequal society that values accumulation, power, and self-interest. They validate a view of humans as driven to competition for dominance; willing to cooperate only with their in-group; and hostile, fearful, and opportunistic toward outgroups unless tamed by "modern" institutions. Alternatively, societies of the deep past may be idealized as living in spontaneous harmony because they purportedly stayed in small kin groups and had no possessions. Neither simplification is appropriate, for both egalitarianism, on the one hand, and social stratification, interpersonal conflict, and intergroup hostility, on the other, depend on sociocultural context.

According to a well-worn categorization of human societies by subsistence strategy, agriculture brought a clean break from foraging and set in motion a process leading relentlessly to social stratification and centralized political control culminating in the modern state. The following discussion calls the bands-tribes-chiefdoms-states progression into question, as well as the assumptions about innate impulses toward hierarchy, dominance, and lethal violence embedded within it. The chapter examines the blurred boundaries between foraging and early agriculture and shows that many large urban settlements functioned without repressive government or administration. It reviews the history of warfare to highlight the

DOI: 10.4324/9781003639923-7

connection between social stratification and intergroup violence and to establish that warfare was not likely to have occurred before the Holocene. The discussion includes an analysis of the characteristics of states, which render inequality intractable and close the door to alternative sociopolitical arrangements. Ideologies of inherent difference such as genetic essentialism serve their purposes by justifying structures of exclusion and unequal access to resources. Current information technologies multiply the reach and impact of mistaken beliefs about the innate characteristics of social groups, highlighting the importance of identifying and contradicting genetic essentialism across educational and popular media.

## Prosocial behavior

Despite the image of humankind as hopelessly self-oriented, people regularly behave in cooperative, other-oriented ways, including showing curiosity toward strangers and openness to social interaction. Studies on wellbeing bear out the antiquity and evolutionary importance of prosocial behavior. Helping behavior and social integration bring better mental and physical health, whereas hierarchy and social isolation have opposite effects, even in situations of economic security and adequate living standards (see Brown et al. 2003; Hodgson et al. 2020; Marmot et al. 2008).

Cooperation emerges early in life together with two related traits: the expectation of fairness and disapproval of cheating. Starting in their second year, babies begin to pay more attention to scenarios in which resources are distributed unfairly than fairly (see Sommerville et al. 2013). Relative to chimpanzees, humans are especially attuned to the need for fairness toward those with whom they cooperate, as illustrated by an experiment in which an unequal amount of food or toys ended up in the hands of one individual in a pair of chimpanzees or 2–3-year-old children, respectively (Hamann et al. 2011). For the lucky chimpanzees, collaboration did not affect how they shared food relative to a windfall or individual-work scenario, whereas the children divided the toys more equitably in the collaboration condition than the other two.[1]

Like adults, children prefer members of their in-group, reflecting the need for trust on the part of a creature born completely dependent on others. However, this preference does not boil down to the ability to recognize relatives. Not only is it likely that over the course of human evolution children grew up mainly in the company of non-relatives (see Chapter 6), but a colored hat or T-shirt is sufficient to engender in-group preference. The strength of in-group preference varies in line with the quality of relationships between groups (see Chapter 1), indicating that it is shaped by context and does not automatically bring fear of strangers or hostility and uncooperativeness toward out-groups. Small children are inclined to help an unfamiliar person even without any reward, as shown by a pair of entirely nonverbal studies with two groups of 19-month-olds (Barragan et al. 2020).[2] Each child watched a man drop a blueberry, strawberry, half grape, or piece of banana (fruits known to be favorites) in four separate trials. In the first experiment, the researcher either dropped the fruit onto a tray opposite the table separating him from the child and ignored it or flopped it over the table and entreated the child to hand it back. Only 1 of 24 children handed the fruit back in the first case (4.17% of 96 trials), whereas 23 of 24 did in the second (34.38% of trials).

To test whether children would hand over the fruit despite being hungry, the researchers repeated the experiment close to children's next meal or snack. This time, 0 of 24 children returned the fruit that was casually dropped onto the tray, whereas 9 of 24 gave it back if the researcher begged (29.17% of trials). Combining the two experiments, 1 of 48 children returned the fruit in the non-begging condition (2.08% of trials), whereas 32 of 48 returned it in the begging condition (31.77% of trials). Among those who returned the fruit, five in each of the two experiments did so every time, despite receiving no sign of gratitude from the researcher, who simply put the food aside.

Experiments with bonobos likewise demonstrate xenophilia rather than other-aversion. In the first of a series of experiments at the Lola Ya sanctuary in the Democratic Republic of the Congo, there were three enclosures with gates in the central area that could be opened only from the inside (Tan and Hare 2013). The test animal would be put in the central area in the morning, before mealtime, a stranger to one side and a familiar animal to the other. The test animals responded by eating a small amount of food placed in their enclosure before opening one of the gates to share the rest. Twelve of 14 did so at least once, in 51 of 70 trials (72.9%). Nine of the animals were more prone to opening the gate for a stranger than a groupmate, whereas three showed the opposite tendency. In addition, once one of the other animals entered the central area, and while there was still food to finish, the third animal was often released, and the one to open the gate was more often the stranger than the test animal. This behavior is significant because it left the stranger outnumbered, a situation which chimpanzees assiduously avoid.

In a second experiment, the test animal had an empty enclosure on one side and a stranger or groupmate on the other. Eleven of the 12 test animals opened the recipient gate at least once (30 out of 60 trials). Again, they were more likely to release a stranger than groupmate, and they opened the recipient gate before the gate to the empty area only if the recipient was a stranger. The third and fourth experiments tested helping behavior where there was no possibility of social/physical interaction. The animals were put in separate rooms from which they could enter mesh tunnels to obtain out-of-reach food. The test animal was left with a new toy, which meant that it would have to sacrifice play time to release the other animal into the tunnels by pulling a rope attached to a key to the recipient's door. Nine of the 10 test animals released the other at least once, yet, in this case, the tendency to release a stranger and groupmate was the same. Finally, in the fourth experiment, the food was placed within reach of the test animal. Releasing the other animal would mean giving it access to food already in the test animal's possession, again with no possibility of physical/social interaction aside from touching through the mesh. None of the seven test animals released the other in any of the trials, indicating that the cost was too high for a behavior that promised no social benefit to the test animal. No aggression of any kind was observed in any of the four experiments, and prosocial behavior did not change over the series of trials, which would have demonstrated the effect of training.[3]

Together the experiments demonstrate strong xenophilia among bonobos, who go out of their way for the opportunity to interact socially with a stranger. The researchers suggest that the extreme other-orientation shared by humans and bonobos originated in selection for high social tolerance, which widens the range of relationships individuals can develop. In humans, the increased xenophilia would have been crucial to the shift to societies in which females and males both migrated out of their natal group, and consequently to the formation of residential groups of mostly unrelated individuals typical of foraging societies.

Cooperative behavior is anything but rare across animal and even plant species. Chimpanzees are more inclined to request a double portion of food to share than a single amount for themselves (see Horner et al. 2011). Pair-bonding, parental-care-sharing prairie voles, but not solitary montane or meadow voles, help each other in times of distress or injury by providing consoling behavior, grooming, and wound care. They experience empathic connection with the distressed or injured individual that is ameliorated by physical contact (see Burkett et al. 2016). Mice will resuscitate an experimentally incapacitated mouse by removing foreign objects from its mouth before proceeding to pull the tongue to clear the airway, which causes the mouse to revive sooner than it would without help (Sun et al. 2025).

Human prosocial behavior extends to care for distant others that the giver will never meet. For instance, people regularly return wallets to their owners and are more likely to do so if the wallets contain money than not. The latter has been demonstrated by an experiment conducted in 40 countries, defying the expectations of both economists and survey respondents (Cohn et al. 2019).[4] Across the 38 countries in which wallets were returned more often if they included money, return rates for wallets with money ranged from 14% to 76%, while those for wallets without money were lower by a roughly constant margin. Countries that landed at the top of the wallet-return scale of civic participation were ones with high levels of social capital and consequently low inequality, such as Sweden and Denmark. A second iteration in three countries increased return rates by raising the level of concern for the owner through greater sums of money and the inclusion of a key.

Similar neurochemical features and mechanisms drive the pursuit of social interaction and its perception as rewarding across social species of marine and terrestrial mammals, including humans. These include a relatively large neocortex at the rear of the cerebral cortex, whose volume varies with the size of residential groups and is especially large in humans. There are the same or analogous neuropeptides, including oxytocin, dopamine, and vasopressin, as well as receptors for them in brain areas involved in social and emotional processes (such as the amygdala and paraventricular nucleus of the hypothalamus), such that defects or disruptions bring aberrant social behavior (see Hung et al. 2017; Sun et al. 2025). The genes that regulate these neuropeptides and their receptors are more alike between humans and bonobos than humans and chimpanzees, such as the vasopressin receptor gene 1a (*Avpr1a*) and the oxytocin receptor gene (*OXTR*) (see Staes et al. 2014).

Engaging in altruistic acts is so psychologically rewarding that it can diminish physical pain, as demonstrated by an experiment in which people who donated blood for the benefit of earthquake victims felt less pain than those whose blood was drawn for routine medical testing, despite the former group giving a greater quantity of blood through a larger needle (Wang et al. 2020). In another experiment by the same team, participants whose arms were progressively constricted by a blood pressure monitor perceived less pain if they were told their contribution would be rewarded with a donation to earthquake victims instead of a monetary reward for themselves. To measure the impact of the meaningfulness of an altruistic act, the researchers had participants rate the helpfulness of a personal financial donation for the benefit of young orphans and then undergo a pain test involving electric shocks to the back of the hand while brain activity was monitored. The more personally meaningful the act, the greater the activation of the ventral medial prefrontal cortex, and thereby the greater the suppression of pain-induced activity in regions including the insula and anterior cingulate cortex. The researchers also examined how altruistic behavior affects chronic pain by engaging hospitalized cancer patients in individual and group activities that either benefited the ward or themselves (cleanup of outdoor areas; preparation of healthful diet plans for the group versus attendance at a nutrition seminar held by a nurse). As expected, while both groups experienced a decline in pain the reduction was greater in the altruism-condition group.

**Primate dominance hierarchies**

The assumption that human biology relentlessly propels people toward self-interested competition is linked to the idea that hierarchy and male dominance are inevitable and prefigured by the behavior of nonhuman primates. Attention to contests for position reflects a traditional emphasis in primatology driven by the interests of scholars from highly stratified, male-dominant societies (see Haraway 2013). Research of the last several decades has shifted the spotlight to female hierarchies; the relationship between coalitional behavior and place in a hierarchy; and prosocial traits, including friendship, play, tool making and use, and emotional intelligence. Basic assumptions have been called into question, including the connection between high rank and better health and reproductive success. For instance, Gesquiere et al. (2025) demonstrate that, for baboons, top position brings increased energetic costs alongside lower psychosocial stress compared to lower positions on the hierarchy, whereas low rank brings reduced energetic costs but higher psychosocial stress. In many species, females and males that form close bonds with kin and non-kin have greater success at reproducing and ensuring their offspring's survival. Examples include communal nesting among ruffed lemurs (Baden et al. 2013) and male bonds among Assamese macaques that correlate with coalitions and, in turn, social dominance and thereby reproductive success (Schülke et al. 2010). Among chimpanzee males, the greatest increment of reproductive success associated with coalitional behavior accrues to those who bring together unallied individuals and consequently those with high social intelligence (Gilby et al. 2013).

Studies on bonobos cast doubt on the centrality of hierarchy and between-group competition to our closest primate relatives. In addition to negotiating looser hierarchies, bonobos eschew conflicts and displays of power, maintain social relationships with other groups, and give each other frequent displays of affection and attachment (see De Waal and Lanting 1997). Females tend to eat first, and males including the alpha defer to them and each other. The outcome of male struggles over rank is decided by females, and, after a scuffle, males smooth things over by mounting each other or rubbing their rumps together. Even among reputedly masculinist chimpanzees, in many groups it is the females that choose the alpha male. Moreover, females mate with all the males during estrus even though higher-ranking males have more opportunities on days of peak fertility.

Cross-species comparisons contradict the assumption that primate males are typically dominant over females. Huchard et al. (2025) analyzed data on male-female contests, which made up about one-half of the total (47.4%) in reports on 253 groups (traditionally and tellingly referred to as "troops") representing 121 species. Quantitative information on the outcome was available for 151 groups from 84 species. If one sex won more than 90% of contests it was classified as dominant. By this criterion, females were dominant in 20 groups from 16 species (13%), males in 25 groups representing 16 species (16%), and neither in 106 groups from 69 species (70%). Of the 106, contests were rare in 21 and won by both females and males in the other 85. Outcomes varied widely between different groups of the same species. For instance, among Patas monkeys, females won contests 0–61% of the time; among bonobos, the range was 48–79%. This variability reflects the fact that dominance behavior is not fixed but rather is responsive to individual characteristics and social and ecological conditions.

Analysis of dominance patterns in relation to physical characteristics reveals that females tend to win more contests in species with no sex difference in the size of the body or canines. Where dimorphism is moderate the more frequent winners can be either sex, whereas males dominate in species in which they are more than 50% heavier than females. Other factors associated with female versus male dominance, respectively, include tree- versus land-dwelling; solitary or paired versus group-living; and monogamous versus multi-mate organization. Human features are a mix: terrestrial and group-living, yet relatively little sexual dimorphism in size or canines, and a range of marriage systems from monogamous to multi-mate.

In short, nonhuman primates do not explain human hierarchies or male dominance, nor could they, for the social, demographic, and ecological environments occupied by humans are unlike those of any nonhuman primate species. The cognitive versatility underlying this range emerged not through violent competition but by way of the many forms of mutual assistance, protection, and social interaction described in previous chapters. Individuals who were willing to collaborate with others were more likely to survive, favoring the spread of traits such as empathy and forbearance. The reduction in aggression opened the door to increased learning. Spoken language and the related boost in improvisation greatly expanded the possibilities for cooperative behavior and the development and propagation of cultural beliefs and practices, far beyond what is possible for nonhuman primates despite their sophisticated vocalizations and nonverbal communication. In sum, if people create gender and social hierarchies, the reason is not buried in distant evolutionary processes but discernable in proximate factors such as those examined below.

## Questioning killer instincts

The ideas about other-danger and male dominance discussed above undergird the assumption that evolution favored violent intergroup competition for territory and females through the reproductive benefit to males with shared genes. Occasional reports of chimpanzee "border patrols" and "warfare" are taken as evidence that these behaviors have deep evolutionary roots. However, the incidents involve unusual circumstances such as exceptionally large groups or rare dynamics of extended internal conflict between opposed coalitions, while violence by multiple animals against a single individual is counted as warfare (see Fry 2006; Mitani et al. 2010). Projections based on human groups are similarly inapt, such as Napoleon Chagnon's warfare-to-reproductive-success connection concerning Yanomami horticulturalists in the rainforests along the Venezuela-Colombia border. Chagnon (1968) argued that murderous men had more wives and children than men who did not engage in killing. Douglas Fry (2006) analyzed Chagnon's strategic and unconventional statistical procedures and found that the relationship did not hold. For instance, Chagnon counted headmen together with killers, even though headmen had more wives and children due to their status; he excluded killers who had died and presumably had fewer children as a result. The key variable was age, for time gave older men more wives and children as well as chances of killing someone. At all events, the study is irrelevant to evolutionary theories about violence given that it does not concern a nomadic foraging society. Moreover, the highly hierarchical and gender-unequal Yanomami are atypical compared to other populations in the region, have endured decades of territorial restriction due to mining and other interests, and may have escalated the violence upon receiving gifts of steel weapons from Chagnon himself.

Lethal violence against conspecifics is anomalous and rare among mammals, suggesting that it has been selected against. Violence carries the risk of reduced survival and therefore reproductive success, which explains why posturing and aggressive displays are far more common. Like other animals, humans engage in far more peaceful and avoidance behavior than violent interactions, and half the species (females) very rarely commits lethal violence.

The biological reasons behind the stronger male propensity for violence as well its variable expression were discussed in Chapter 5. Cultural contexts may favor other-orientation, restraint, and mutual respect, or their opposites, and personal characteristics and life experiences differ across individuals. Throughout the world, violence tends to be viewed negatively, while people who relish killing are uncommon and considered deviant and dangerous. They may be subjected to capital punishment by nations and indigenous societies alike and are rejected for military service (see Fry 2006). To help soldiers overcome the human aversion to killing, militaries employ psychological techniques that strengthen feelings of duty, belonging, and fear of noncompliance. Psychoactive drugs may be needed to face modern combat in which it is difficult to distinguish between enemy forces and noncombatants, and technology magnifies the scope of killing. The harm to the self that comes from hurting others is clear in high rates of psychological disorders and suicide among soldiers upon return to civilian life, as well as among commanders who experience moral injury from having made decisions they knew would cost lives (see Grossman 2014). Post-conflict interventions affirm the contribution of social connectedness to wellbeing, for programs that emphasize sustained social interaction in culturally appropriate forms such as mentoring networks, political organizations, or games and team sports consistently bring positive impacts, whereas interventions less focused on building relationships – such as formal truth-and-reconciliation meetings – may aggravate psychological distress (see Blessin et al. 2022; Krasniqi and Krasniqi 2019).

That killing does not come easily to humans makes it curious that warfare is widely described as the norm across all time. Textbooks and scientific literature depict menacing Stone Age males wielding weapons and, until the discovery of interbreeding, implied that humans hunted down and killed other species, including Neandertals (see Ghosh 2022). These other species did not need humans to wipe them out, for their own characteristics and demographic patterns were sufficient (see Chapter 3). Besides interbreeding, the exchange of knowledge and technology indicates friendly rather than hostile interactions. For instance, 47,000-year-old humanmade artifacts in a cave in Bulgaria (Bacho Kiro) used by Neandertals thousands of years earlier include utilitarian and decorative objects such as pendants made from cave-bear teeth which resemble those attributed to Neandertals 3,000 to 2,000 years later at sites such as the Grotte du Renne in France (Hublin et al. 2020). Contrariwise, Neandertals created sophisticated objects and art before encountering humans that are likely to have flowed in the opposite direction, although in all cases independent invention cannot be ruled out.

Throughout the Pleistocene, group sizes were too small and resources too openly available for foragers to be territorial or maintain a class of warriors. At the tail end of the Pleistocene, there were 0.3 people per 100 square kilometers. This was an increase from 0.2 around 25,000 BP and 0.07 circa 300,000 BP. The Holocene brought population density to around 34, or more than 100 times greater, which is not far from a worldwide sample of recent and contemporary foragers (all continents except Europe): 12.3 among 21 mobile foraging bands, 56.9 among 9 complex foragers, and 25.7 for all 30 together (Fry et al. 2020). Warfare is rare or nonexistent among societies in the mobile foraging sample (see below) despite high population density relative to the Pleistocene. As noted in the last chapter, instead of fighting Pleistocene foragers were more likely to have shared food, cared for one another, and met up for friendly mass gatherings.

The two earliest cases of apparent warfare involve semi-settled or settled societies in areas alongside bodies of water, at a time in which climate changes had reduced the size of animal populations and may have generated competition over aquatic resources: Jebel Sahaba, near inlets to the Nile River (11,600 BP) and Nataruk on the former shores of Lake Turkana (10,500 to 9,500 BP) (Fry et al. 2020). In the first case, a cemetery indicates long-term riverside residence and therefore a society unrepresentative of nomadic foragers. Initially, 24 of 61 skeletons were thought to show signs of violent death from projectile weapons. Subsequent analyses suggest other causes and leave only a handful of skeletons with unambiguous evidence of lethal aggression (Fry et al. 2020). In addition, the existence of both healed and non-healed lesions on the same skeletons indicates lesions endured over time rather than all at once, whether through interpersonal violence or intergroup conflicts (see Crevecoeur et al. 2021).

At the Nataruk site, pottery suggests settlement and food storage on the shores of a lake much larger than its current size. The partial remains of at least 15 individuals do not show signs of violence, whereas damage in 10 of 12 complete skeletons appears to indicate a massacre. However, it is not clear that all 10 died at the same time. The embedded stone blade in one skeleton could have caused nonlethal injuries and/or been accidental. In addition, the condition of the skulls does not point to a lethal blow to the head. An analysis of data on 400 sites and around 3,000 individuals from before 10,000 BP found no additional evidence of pre-Holocene warfare aside from one site with three individuals buried together although with no sign that they died violently, and four sites with one or more skeletons whose bones were pierced by projectile points, none of which firmly implicates warfare instead of accidents or interpersonal violence (Haas and Piscitelli 2013).

Interpreting skeletal damage is often complicated because the marks could be the result of natural forces such as compression by rocks and soil, teeth marks made by animals, and the work of insects. The ambiguity allows for the coexistence of opposite views of warfare as having deep or shallow roots in human history. While the reality may lie in between (see Meijer 2024), recent and contemporary foraging societies cast doubt on the deep-roots perspective. Unlike prehistoric foraging groups, ethnographic groups of foragers have faced incursions by antagonistic outsiders (colonists, missionaries, governments, farmers), territorial restriction, forced displacement, and the introduction of trade goods, metal weapons, and addictive substances. These conditions have no parallel in past societies (see Fry and Söderberg 2013; Fry et al. 2020). Moreover, the intergroup violence among nomadic foragers is unlike that of settled and equestrian foragers or village-based horticulturalists such as the Yanomami which are held up as exemplars of primordial killer instincts. The deep-roots perspective draws upon studies that classify all violent deaths in small-scale societies as warfare, even if there was a single victim and even if the cause was capital punishment, accidental death of a bystander or peace-maker, sorcery, feuding, or revenge killing.

Analysis of 30 ethnographic foraging societies from the Standard Cross-Cultural Sample reveals a similar amount of lethal interpersonal violence between mobile ($n$ = 21) and settled ($n$ = 9) societies (Fry and Söderberg 2013; Fry et al. 2020). These include one-on-one violence, more than one perpetrator and one victim, and one perpetrator and more than one victim, averaging around 4–5 events in total. In contrast, events involving multiple perpetrators and multiple victims are far more common among settled societies than mobile ones: an average of 7.8 versus 1.4 events. Thanks to this difference, settled societies have a higher overall total of lethal events: 12.1 versus 7.0.

Evidence from the mobile forager sample echoes the archeological record of occasional lethal interpersonal violence without warfare. Most of the lethal violence in the sample occurred between two people, with the rest split between more than one perpetrator and a single victim and more than one perpetrator and more than one victim. In only 1 of 148 cases did a single person kill more than one victim (two children). Women were killers or co-killers in 4% of cases. One society, the Tiwi, accounted for 69 lethal events, compared to 79 for all others combined. Among the other 20 societies, in 3 there were no lethal events and in 7 there were 3 or fewer. In 10 societies, there were no events involving more than one perpetrator.

The Tiwi case is unusual both for the number of events and their type. Tiwi events were less likely to occur within a group (35% versus 63% among the other 20 societies) and more likely to occur between groups (55% versus 15%). Three-quarters of all intergroup events involved the Tiwi (38 of 50). None of the Tiwi events concerned contested resources whereas two of the others did. While interclan revenge-seeking accounted for nearly half of the Tiwi intergroup conflicts it explained none of them in the other societies. Among the other 20 societies, 5 of a total of 12 events involving lethal violence against outsiders were committed by the Hadza in conflicts with neighboring cattle herders (not foragers). These patterns indicate that lethal violence in nomadic groups other than the Tiwi rarely involves outsiders or more than one perpetrator and is driven mainly by personal or interfamilial motives.

The much higher frequency of multiple-perpetrator, multiple-victim events among settled nonegalitarian foragers reflects their different social and demographic characteristics compared to the mobile foraging sample (Fry et al. 2020). Eight of nine North American societies in the settled sample had class distinctions (two hereditary, six class-based). Group size was double that of nomadic populations, while temporary aggregates were half as large. Settled societies depended heavily on fishing and moved six times less frequently per year (2.0 versus 12.3). Population density was 56.9 per 100 kilometers, or more than 4 times greater, and there was greater population pressure (food availability divided by population density). Settled societies were more likely to take slaves and have leaders with authority and privilege.[5] High values for these factors correlated with warfare among the societies in the sample, whereas the characteristics of mobile egalitarian societies explain the complete lack of lethal events in some societies and the near absence of anything that could qualify as warfare in the entire sample. The difference between the two suggests that the emergence of inequality and warfare is question of history, not inescapable biological or evolutionary destiny.

## Egalitarianism

The above has shown that the conditions that engender warfare are not typical of mobile foraging societies and even less likely for those of the deep past. While egalitarianism clearly militates against intergroup violence, it does not entirely prevent interpersonal violence. The following discussion explores factors which sustain egalitarianism and highlights how it is neither spontaneous nor absolute.

Egalitarianism promotes an even distribution of valued resources such as material goods, knowledge, access to the spirit world, and a say in group decisions. It precludes the ability of individuals to issue orders to one another. However, egalitarianism does not eliminate differentiation or discord. People in foraging societies have variable amounts and types of personal property, including equipment, containers, and jewelry. Some people are more talented, physically attractive, or charismatic than others. Women and men may be admired as spiritual leaders or especially productive hunters and may take on a leadership role, although their ability to direct others depends on persuasion and example rather than birthright or institutionalized position. Disputes arise as they do in any group of people, and over the same main issues: partnership/sexual conflicts and failures to behave fairly, as in unmet obligations or inadequate contributions to the family or group (for specific incident counts, their reasons, and their resolution in various foraging societies, see Montagu 1978). Fights occur and sometimes there is a homicide. More often, restraint puts a damper on conflicts, while those which burst out are resolved by the intervention of others before serious harm is done.

Factors that contribute to egalitarianism and peaceful interactions within and between foraging groups include large territories and beliefs that curtail overexploitation of collective resources, such as custodianship of watering holes held in trust and allocated upon request in southern

Africa, or totemic systems that protect animals associated with clans in Australia and North America. The fluid, bilateral kinship organization of foraging societies disfavors intergroup hostility given the dispersal of relatives across groups, while the freedom to change residence keeps internal tensions from escalating. Mobility limits storage, precluding wealth acquisition as a motivation for raiding. Violent confrontations over women (or men) concern specific individuals, not people the fighters do not know. The system defends both interdependence and individual autonomy through food sharing and gift exchange, leveling mechanisms such as taboos against using valued meat strategically, a lack of formal leaders and political institutions, consensus decision-making, collective childrearing with strong male involvement, and cultural beliefs that discourage self-aggrandizement, greed, arrogance, and power.

Even societies of settled foragers offset the concentration of wealth and power through reciprocity and redistribution (see Mauss 1990; Sahlins 2013). For instance, the potlatches held by Salish and Kwāg'ul (Kwakiutl) lineage chiefs on the Northwest Coast of North America involved feasts and gifts of stockpiled goods, including copper plates and wool blankets, possibly thrown into the sea. Such practices buffer against economic insecurity and inhibit the growth of internal inequalities. Prestige comes from giving away wealth rather than accumulating it.

Settlement encourages unilateral kinship organization, which clusters relatives and promotes competition between kin groups. Food and water storage allows for the complexification process involving the emergence of social classes, gender inequality, prestige trade, formal leadership, and defense of territory. Over time, warfare may become endemic where the process generates a shift from reliance on defensive landforms to the construction of defensive structures and earthworks (Fry et al. 2020).[6] Ancient examples include Holocene foragers in the Americas, Europe, the Near East, Japan, Eastern China, and the Far North. Ethnographic cases include settled foragers in Florida, the Northwest Coast of North America, Alaska, Kodiak Island, and New Guinea. In New Guinea, intergroup violence peaks among fisher-foragers in dense, populous, sociopolitically complex settlements, compared to semi-settled foragers inhabiting freshwater wetlands. Contrariwise, historical Inuit societies sharing the same type of rich aquatic environment did not uniformly take the path of complexification and increased warfare: groups of Iñupiat in Northwest Alaska regularly engaged in warfare, whereas another 12 Inuit societies in Canada and Greenland very rarely did so. Instead of honoring violent men and training young men to fight, the latter discouraged killers and valued skill in hunting (Fry et al. 2020).

Ethnographic societies that practice (or practiced) a mixture of horticulture and foraging in warm, humid parts of the planet illustrate how settlement need not but can be associated with complexification and warfare. These societies generally occupy warm, humid, resource-rich parts of the planet, including areas of Africa, South Asia, Melanesia, and the Americas, where the climate does not permit food storage. The lack of storage limits economic inequality and prevents the accumulation of surpluses that can be appropriated as tax or tribute by kingdoms or states, as occurs in cooler, drier places where farmers grow grain. Villages are moved every few years or decades, especially where communities practice slash-and-burn cultivation. The use of simple, easily accessible tools such as hoes and digging sticks keeps food production in the hands of everyone, with tasks divided or shared in numerous ways according to gender and age. Kinship may be matrilineal, patrilineal, or bilateral. Leaders may be women or men who ensure fairness, resolve disputes, and provide care, but whose authority and position hinge on example and persuasive oratory. Redistributive feasting and gift-giving, reciprocal gift exchanges, trade, and marriage relationships between groups contribute to peace, whereas disruptions to these practices bring increased intergroup conflict. Warfare is infrequent relative to agricultural societies. It is more common among societies with patrilineal kinship and more pronounced gender inequality manifested in reduced female responsibility for production and other valued tasks, separate eating and sleeping areas for women and men, and beliefs that construe women's bodies as dangerous and polluting (see Nolan 2003; Whiting 1975).

Scores of internally peaceful societies with no history of warfare span all the major world regions (see Fry 2006). Together with the egalitarian Holocene societies discussed below, they stand as evidence that political centralization, stratification, and structural inequality are not inevitable for human societies.

## Seeds of stratification

As noted above, gender inequality is central to complexification and correlates with warfare. The last chapter showed that the earliest division between female and male subsistence and related tasks can be traced to the Late Pleistocene when the size of foraging groups increased sufficiently. This differentiation planted the seed for reduced egalitarianism, yet there is a limit to inequality where everyone has access to the necessities of life.

Among ethnographic foraging societies, gender egalitarianism declines in step with increases in men's control over the distribution of meat. Meat from large animals is unpredictable, high-quality, and shared beyond the family, which gives it exchange value that can be used to create obligations (see Zeanah et al. 2026). In contrast, food shared within the family has use value. Ernestine Friedl (1978) analogizes forager women's collection of food for their families to paid workers' spending their incomes on family expenses in a cash economy. In both cases, the product of women's labor does not have the strategic value of meat or money used by men in the wider society. Friedl observed the most egalitarianism among societies in which teamwork was the rule, as among the Agta of the Philippines or Mbuti, Efe, and Aka foragers in central Africa who practice net hunting in groups. After them came societies in which women and men worked independently, such as the Hadza of Tanzania, followed by those with interdependent roles such as the Ju/'hoansi of the Kalahari Desert or the Tiwi of Melville and Bathurst Islands off the Australian coast. While still relatively egalitarian, among the Ju/'hoansi productive hunters influenced collective decisions and attracted more people to their group; Tiwi men could have more than one wife. Last of all, and off the charts, were the Inuit because the men procured nearly all the food and there was female infanticide. However, some customs may not have been as they appeared, such as wife stealing that was consensual, while conflicts over women could be resolved through ritual fighting and song duels (see Fry 2006; Hoebel 1954). Moreover, as noted in the last chapter, women in arctic environments procure other foods and provide essential support by transporting and processing large game.

Another mechanism associated with reduced gender egalitarianism concerns food and water storage, which also contributes to the emergence of social classes. While other San populations of the Kalahari distributed meat immediately and carried only small amounts of water in ostrich eggs, G//ana foragers kept goats and stored dried food at a year-round site, including meat and maize, melons, and cultivated legumes (Cashdan 1980). Some of the men earned money working in mines outside the reserve. These activities allowed people to buy water-storage containers and, in some cases, horses, donkeys, and cattle kept by relatives outside the reserve. Wealth objects such as furs and skins were owned by 32 out of 105 people. One-fourth of the men had sufficient livestock to pay bridewealth for two or three wives. Wealthy men acted as leaders of the community and were respected rather than ridiculed for behaving in greedy or self-advancing ways. Even though people continued to share wild and cultivated food, accumulation was culturally acceptable.

These features of G//ana society explain why semi-settled or settled foragers take to herding or farming more readily than mobile foragers. James Woodburn, who studied among the Hadza in Tanzania from 1957 to 1960 and over many visits thereafter, explained that the Hadza kept an immediate-return system that involved sharing food the day it was collected; the right to work when and how one wished; an ethic of individual autonomy; and social equality, especially between generations (unlike herding societies, in which only male heads of household are equal) (Widlock 2002).

Kinship and residential organization were flexible. Given that foraging takes relatively little time and consequently is not people's main activity, there was abundant time for gambling for arrows and moving camp frequently. Unlike their agro-pastoral neighbors, the Hadza did not have experts, leaders, or litigators and did not value oratory or the ability to deliver an argument.

Woodburn contrasted the Hadza system to the delayed-return system of Okiek foragers in Kenya, which was rooted in a honey economy that involved equipment and territorial management. Like indigenous American societies of the Northwest Coast that had traps, boats, and weirs, the kinship system included clans, lineages, and binding obligations between relatives. The delayed-return system created institutionalized interdependencies that allowed the Okiek to transition effectively to farming and herding, whereas the Hadza and other foraging populations in Africa struggled to become farmers and accept village life even when they tried voluntarily.

The immediate-return system has persisted among the Hadza and other foraging societies to the present day, despite their declining dependence on foraging and increasing contact with neighboring villagers and tourists. Paradoxically, the system is both resilient and a vulnerability. Mobility and egalitarianism limit political organization, opening nomadic foraging societies to further losses of territory and resources through government appropriation, farmers who clear forests to plant crops, and herders who reduce the supply of water and game by pasturing animals, digging wells, and building fences.

A mixed-methods analysis confirms that the historical reason nomadic foragers have been confined to ever smaller territories is the emergence of stratification (Rogers et al. 2011). The analysis included computer simulations, 15 archeological case studies, and historical accounts of early Polynesian agricultural societies. It demonstrates that class-based societies outcompete egalitarian ones, whether in a constant environment or one with fluctuating resource availability. Their greater readiness to migrate explains why class-based farming and herding societies filled the space between nomadic foragers and the egalitarian Holocene settlements described below and eventually covered nearly all the Earth's landmass.

## Complexification averted

The term, Neolithic Revolution, implies that agriculture brought a clean break with the past. Conventional wisdom holds that this fatal and decisive step caused progressively greater political complexity leading to the state and its centralized and professionalized politico-governmental organization supported by a permanent cadre of administrators, soldiers, and police empowered to use violence on its behalf. Along the way, complexified societies engaged in endemic violent competition with other societies over natural and human resources unevenly distributed across the planet. In compensation for the bloodshed, misery, and inequality, complexification is credited with generating specialized classes of scholars, artists, and religious experts, and all their accomplishments. Once the process began, there was no stopping it.

This storyline misrepresents both the timing of many human achievements (see below and Chapter 6) and the way agriculture emerged and spread. There were continuities between foraging and early farming societies, while urbanization, farming, and hierarchy did not always go together. Egalitarian urban settlements with tens of thousands of residents preceded full commitment to agriculture. They engaged in public works projects, built uniform housing for all, and redistributed surpluses and land equitably, as shown below.

In other words, agriculture was not a monolithic force that crushed foraging societies underfoot and automatically generated hierarchical ones. Genomic and linguistic analysis of the diffusion of the Indo-European, Bantu, Austronesian, and indigenous American language families reveals a variable and temporally uneven process of farmer demographic/linguistic expansion (Stoneking et al. 2023). In some areas, the diffusion of genes and languages was not related to the spread of agriculture; in others, demographic, agricultural, and linguistic expansion went hand in hand, yet the genetic impact of contact between residents and incoming farmers ranged from admixture to replacement. In short, numerous outcomes of intergroup interactions were possible.

As seen in earlier chapters, settlement began long before the Holocene. There were seaside and other waterfront villages in the Pleistocene, and hundreds or thousands of foragers would co-reside for extended periods to trade goods, build monumental art and buildings, and take advantage of seasonally resource-rich environments. These temporary aggregations continued in the Holocene, when a warming trend brought the retreat of glaciers and the filling of lakes, estuaries, and waterways, as well as expanded forests with diverse edible plants and nuts along with smaller game.[7] Leadership and administrative organization were needed to manage ceremonial events, resource utilization, crowded settlements, and monument construction, yet this temporary political complexification did not bring permanent settlements or engender lethal confrontations. Similar periodic agglomerations are typical of recent and contemporary foragers in Africa, on the Northwest Coast and Great Plains of North America, and in South America.

The maintenance of egalitarianism in huge settlements of foragers and the early agriculturalists described below calls into question longstanding dogma about the inevitability of complexification where numbers exceed

150 (Dunbar 2003). This limit corresponds to the presumed maximum number of people with whom ancestral foragers interacted as well as the brain's limit for retaining personal information about individuals. That is, larger populations are thought to present excessive cognitive demands to a brain molded by small-group nomadic life. Not only does the argument rest on the mistaken presumption that people in small groups automatically get along better than those in large ones, but past foragers met in large numbers while contemporary foragers describe knowing far more people than the alleged limit (see Bird et al. 2019; Hill et al. 2014).

Just as settlement preceded agriculture, so did purposeful environmental management and familiarity with domesticable plants and animals. As long ago as 50,000 BP or earlier, humans transported seeds to new areas, cleared forests with fire, and cut down trees to open the understory for plant growth.[8] By 20,000 BP wild animals were trapped and moved into less populated areas. The period from 13,000 to 12,000 BP brought seed preservation, grinding stones, and the exploitation of more productive environments through the introduction of plants and animals. During the Early (Greenlandian) Holocene, beginning 11,700 BP, the deglaciation that created new productive environments mentioned above also brought local changes in climate and rainfall that rendered the supply of edible wild plants and game less predictable. Humans responded by domesticating barley, rice, wheat, and rye, followed by sorghum, millet, and other grains. In the Americas, they domesticated squash, beans, potatoes, maize, and peppers. Farming could support larger communities by increasing the ratio of calories to unit of land and, in turn, fueled faster population growth by way of soft supplemental foods that hastened weaning and reduced interbirth intervals.

In the Early Holocene, animals such as sheep and goats were captured to fatten as a form of mobile storage. By the first centuries of the Middle (Northgrippian) Holocene 8,300 BP, these and other animals had been domesticated, including boar and cattle. The Middle Holocene began with a cooling trend that led to the calming of the world's great rivers, which created especially fertile soil in broad floodplains. Some settlements mixed farming with foraging and maintained a varied diet and relatively sustainable relationship with the natural environment; others which cultivated cereals extensively had a more monotonous diet and were more vulnerable to crop and livestock losses. As stratified societies the latter grew and spread more readily (see above), despite heavier work and higher mortality (see Cohen 1989). The two-century drought that marks the start of the Late (Meghalayan) Holocene 4,200 BP created severe challenges for large, cereal-dependent dynastic and imperial societies, whose histories of warfare and internal strife are well known. Holocene subdivisions are outlined in Table 7.1.

*Table 7.1* Dating of Early, Middle, and Late Holocene

| | *Date Range* | *Marked by* |
|---|---|---|
| Early Holocene | 11,700–8,300 BP | Deglaciation |
| Middle Holocene | 8,300–4,200 BP | Cooling trend |
| Late Holocene | 4,200–Anthropocene | Drought |

The present discussion focuses on the Early and Middle Holocene, a time in which large settlements combined agriculture, urbanization, political organization, and administration and technology in multiple and reversible ways. The communities that emerged in this period indicate that there was no unidirectional demographic/political moving sidewalk leading to stratification and centralized politico-administrative-military power and control.

Both in recent times and prehistory (see Chapter 6), foragers have shown a strong tendency to reject farming despite familiarity with it from contact with neighboring villagers. People taken away from or persuaded to leave their indigenous society to join a "modern civilization" nearly always have escaped back to the bush if they could (Graeber and Wengrow 2021). This reaction suggests that early agriculturalists were not the envy of foragers. Holocene foragers would have had deeper knowledge of the ecosystem. They were better off materially: the goods left behind by the local farming population were no match for the monumental constructions, trade centers, and lavish burials left in locations such as Göbekli Tepe. Greater wealth and power may have allowed Holocene foragers to demand tribute from farmers or take them as servants, as 16th-century Europeans observed among Guaicurú foragers vis-à-vis Arawakan villagers in Paraguay (Graeber and Wengrow 2021).

Many early farming communities in both the lowland (Jordan, Tigris, Euphrates River valleys) and northern upland areas of the Fertile Crescent, including very populous ones, did not immediately take fully to agriculture but rather combined it with foraging or shifted between the two over hundreds to thousands of years (see Graeber and Wengrow 2021). In Central Anatolia, the built environment of farmers/herders/foragers consisted of beehive-like collections of mudbrick houses with openings at the top and underfloor tombs for family members. The size and quality of houses did not differ, nor were there gender differences in burial customs or diet (as indicated by osteological analysis).

One such settlement, Çatalhöyük, was occupied by up to 8,000 people from 9,500 to 7,600 BP. Genomic analysis of the skeletal remains of 131 individuals from 35 houses dated between 9,000 and 8,000 BP indicates that family groups were matrilineal, with 70–100% of girls remaining in their natal households while boys migrated out (Yüncü et al. 2025). At the beginning, burials included cousins, aunts, and nieces, indicating extended family groups. Over time, the people buried together came to be less genetically homogeneous, suggesting a weakening of the importance of kin relationships as the core of social organization. Female-centered values and practices prevailed over the entire period, nonetheless. Female infants and children were buried with five times as many grave goods as males. The site's abundant female statuettes suggest goddess worship and fertility themes; many show women in poses of authority, indicating high politico-religious status and roles. Similar cultural practices have been found throughout the region, and beginning 1,000 years earlier, reflecting a broad tendency toward female-centered kinship and community organization.

To the west of Çatalhöyük around 9,500 BP, populations genetically continuous with foragers living in the area 5,000 years earlier shared diverse collections of cultural traits with Neolithic agricultural societies of the Fertile Crescent, including Central Anatolia, but it was not until afterward that they began to integrate genetically with incoming farmers from the east (Koptekin et al. 2025). An analysis of 30 new paleogenomes from sites including Çatalhöyük dated to 9,800–8,000 BP, hundreds of published ancient genomes, and 54 elements of material culture from 16 sites found that none of the genomes from after 8,500 BP resembled the earlier population, suggesting that it had integrated with the newcomers or disappeared. Over a few hundred years, an "Anatolian farmer" genetic profile was established. Migrants with this profile then spread across the Aegean and throughout Europe. The analysis established that relatives were dispersed across wide areas of Anatolia and the Aegean, and that geographical proximity rather than genetic similarities explained the distribution of cultural traits. That is, culture spread between villages independently of and much more swiftly than genes.

Because of matrilocality, most of the outmigrants from Anatolia are likely to have been males. Over time they replaced female-centered institutions with patrilocal, patrilineal kinship and increased male status reflected in lavish graves for male elites. Gender and class differences emerged at the household level, as shown by homestead cemeteries in the Lech Valley, Germany, 4,750 to 3,300 BP (Mittnik et al. 2019). Households were made up of high-status males related to others in the community; high-status in-migrant females; and low-status females and males unrelated to other individuals in the household. High-status males were buried with both ornaments and implements, high-status females with ornaments only, and low-status individuals without rich grave goods. That status was inherited is evident from a site in which the only 3 individuals (out of 16) with rich grave goods were a mother and two sons.

Alongside the unequal, warlike dynasties and empires that dominate the historical record were Middle and Late Holocene societies that farmed or mixed foraging and farming for hundreds to more than a thousand years without signs of stratification and inequality (Graeber and Wengrow 2021). These settlements preceded dynastic polities such as Uruk, Harappa and Mohenjo-daro, Shang, Inca, and Maya, and persisted for hundreds of years thereafter in Mesopotamia, the Indus Valley, China, and South and Central America, respectively. Urban settlements could have tens of thousands of residents, or even more than 100,000 in Teotihuacan ca. 2,100 BP. Some societies abandoned settlements to return to foraging; others inhabited urban settlements periodically, only to disperse in between. Houses did not differ in size or quality. In some cases, they were densely packed and encircled by farmland and forests. In others, each dwelling had a kitchen garden. Societies expanded and regulated themselves without extensive agriculture. They produced surpluses and constructed public works such as irrigation systems, public plazas, and ceremonial centers where a ritual calendar celebrated civic membership in a collective united by shared divinities or founding ancestors.

Egalitarian Holocene societies managed complex tasks without repressive and exploitative politico-administrative apparatuses (Graeber and Wengrow 2021). They resolved disputes, maintained public safety, built infrastructure, organized imports and exports, cared for the needy and sick, and distributed economic resources and responsibilities. Some societies reverted to more egalitarian social and political arrangements after a period of elite housing and graves, defensive walls, monument building, ritual killing, and other manifestations of inequality and centralized authority. For instance, residents of Mesoamerican Teotihuacan around 1,700 BP and Taosi in China around 4,000 BP destroyed monuments and palatial buildings, equalized housing, and created art with same-sized women and men and without images of rulers dominating subordinates. In other cases, societies abandoned centralized order for smaller republics with a lighter environmental impact, as in Cahokia and successor kingdoms in the North American Midwest beginning around 1350. Elsewhere yet, egalitarian housing and grave goods gave way to inequalities, defensive earthworks, and warfare, as in Austria and Germany between 7,500 and 6,500 BP.

Changes in inequality in the Carpathian Basin from 8,000 to 3,000 BP were minor, variable, and short-lived, as shown by an analysis of burials and house sizes at over 100 sites (Duffy et al. 2025). Leveling mechanisms that kept inequality from taking root and persisting across generations included collective labor on infrastructural projects, the occasional splitting of communities, and the burial of wealth objects. Villages that became a huge city in today's Ukraine and Moldova were settled by migrants who arrived by way of the Carpathian Mountains and inhabited for 800 years starting around 6,100 BP (Graeber and Wengrow 2021). Robust houses of equal dimensions, each with a kitchen garden, were built in a circular

**Characteristics of egalitarian settlements**

- surpluses redistributed; wealth given away or buried forever
- low-impact farming and foraging
- no social class or sex differences in burial goods and/or artistic representation
- houses of equal dimensions

*Figure 7.1* Holocene egalitarian settlements.

tree-ring pattern around an enormous central area where animals may have been pastured or public gatherings held. There was a complex economy that generated surpluses, yet no sign of centralized storage, public administration, monuments, or defensive structures.

While huge early settlements functioned without technologies associated with "civilization," such as horses, advanced metallurgy, or complex accounting systems, they generated many of the innovations attributed to it, including arithmetic and geometry, writing, long-distance trade, and diplomacy (Graeber and Wengrow 2021).[9] Knowledge was developed long before being put to practical use. In Mesoamerica before the arrival of Europeans, toys for children included wheels complete with axles and spokes, yet there were no wheeled vehicles. Similarly, the Upper Paleolithic figurines discussed in the last chapter preceded the use of clay for pots and storage containers by thousands of years. Moreover, early symbolic systems for keeping count appear to have been devised not for the sake of tax and tribute but to ensure fairness in the distribution of stored goods, the payment of debts, and the making of contributions (including labor) to collective projects such as religious or public ceremonies and buildings. For instance, in the Mesopotamian village, Tell Sabi Abyad, geometrically shaped clay tokens were made 8,000 BP, three millennia before the invention of writing. The lack of differentiation among the houses suggests that the tokens were used to allocate resources kept in communal granaries and storage facilities, although, as in all cases, caution is needed in interpreting archeological materials. Some of the characteristics of egalitarian settlements are summarized in Figure 7.1.

## Political centralization

Given that things could have gone differently, David Graeber and David Wengrow (2021) ask how and why much of the global population has gotten "stuck" in stratified societies with centralized political authority. Based on detailed examples from 5,000 to 4,000 BP, their proposed answer is that states join three forms of domination whereas kingdoms and chiefdoms manage only one or two: sovereignty, or control over violence; bureaucracy, or control over knowledge/information including esoteric or official secrets (whether sacred or secular); and heroic power negotiated through competition among charismatic elites. First-order polities such as the Olmec and Chavín de Huántar imposed only one: charismatic competitive politics and bureaucracy, respectively. Second-order polities combined two: sovereignty and bureaucracy in Old Kingdom Egypt; bureaucracy and charismatic power in Mesopotamia; and sovereignty and charismatic power by the Classic Maya. In each of these second-order polities, the missing principle was sublimated into the sphere of divinities and the dead: competitive politics in Egypt; sovereignty in Mesopotamia; bureaucracy among the Classic Maya.

Third-order polities (states) combine the three forms of domination under the precept that domination originates in and serves the people: power derives from the people; bureaucracy exists to protect and benefit them; and political competition is handled by their representatives. The three forms of domination stifle three freedoms protected by egalitarian societies: the freedom to leave/move; the freedom to disobey; and the freedom to create and revise sociopolitical organization. Some freedoms persist where the three forms of domination fail to converge. For instance, a kingdom or empire may enforce compliance with its rules but be unable to stop people from leaving.

Because of their inability to unite all three kinds of domination, kingdoms were fragile. This is why, in the process of establishing themselves, they often engaged in monument building and dramatic mass violence, such as burials with tens to hundreds of sacrificed people and animals. Their paramount rulers were put through rituals involving physical ordeals to encourage humility, and overly autocratic leaders could be replaced. "Big men," chiefs, pharaohs, and monarchs typically were men but could be women, chosen by women, or granted their position through matrilineal kinship ties. As shown below, it is states that systematically marginalize women from politics and religion.

A few examples illustrate the political importance of women in nonstate societies. One is the first wide-ranging political structure north of the Alps, spanning from Iberia to Eastern Europe. Its "early Celtic" chieftains obtained their rank matrilineally, as indicated by 31 genomes and 3 elite burials from 2,600 to 2,200 BP (Gretzinger et al. 2024). Among many indigenous societies of North America male leaders were chosen by women or designated by matrilineal ties. In others, leaders were women or both women and men. The extensive Hopewell area of the midwestern US from 2,100 to 1,500 BP provides an illustration. Villagers built enormous earthworks and secreted away great stores of treasures in burial mounds, despite modest farming that did not yield large surpluses. In the southern sites all burials that contained badges of office were females; in the northern sites, males; and in the center, a mix (Graeber and Wengrow 2021).

Despite being surrounded by and in regular contact with male-dominated polities in Egypt, Mesopotamia, and Anatolia, the island of Crete appears to have been female-centered and egalitarian from 3,700 to 3,450 BP (Graeber and Wengrow 2021). Settlements such as Knossos and Phaestos had tens of thousands of residents, surpluses and storage depots, palatial buildings, and a system of writing, yet appear not to have been led by a supreme ruler. Minoan visual imagery depicted women sitting on thrones, conducting fertility rituals, appearing beside dangerous animals and mythical beings, and attending all-women assemblies. Women were always clothed, whereas men generally were either naked or barely covered and portrayed as athletes or supplicants bringing tribute to women officials. Women were drawn at larger scale than men, signifying higher political status. The trade conducted with other Mediterranean societies had a decidedly feminine orientation.

Alongside the chiefdoms, kingdoms, and empires that emerged throughout the world in the Middle and Late Holocene were self-governing communities of the sort described above (Graeber and Wengrow 2021). Some were absorbed by competing dynastic polities on their outskirts, as in the case of Mesopotamian city-states. Invading rulers had to contend with local systems of governance and responded by incorporating some of their customs, such as seasonal corvée labor expected of all people including elites in Egypt, the Near East, and Eurasia. Widespread traditions prescribing hospitality to newcomers, including strangers, continued to facilitate people's ability to leave their society for another.

In contrast to first- or second-order polities, states give people limited scope for leaving or disobeying and, instead of allowing reflection on or experimentation with different forms of sociopolitical organization, they confine them to one. While the process of state formation is beyond the scope of the present discussion, a common thread involves permanent, year-round, extensive cereal-based agriculture, which correlates with social stratification, sociopolitical complexification, and increased internal and external violence. Cereal agriculture, and particularly plough-based farming, is tightly associated with patrilineal kinship and gender stratification, a relationship evident today in the unequal gender beliefs of societies where the plough is or was used in cereal cultivation (see Alesina et al. 2013). Grave goods and skeletal damage in burials from the period 6,000–5,000 BP point to the emergence of gender-divided tasks in early cereal-based farming communities in central Europe and elsewhere, although with variability within sexes and overlaps between them: greater male participation in hunting; female labor pounding grain with heavy stones; increased amounts of ornaments and tools in male than female burials; and violence against females and children (see Lacy and Ocobock 2024; Masclans Latorre et al. 2021).

Patrilineal kinship concentrates power and resources in the hands of men, interrupting the cross-cutting interdependencies spun by matrilineal and bilateral systems and isolating women from their birth families. Men gain control over land, equipment, yields, wealth, trade, politics, law, religion, and the behavior and material wellbeing of women and children, including women's sexual freedom and access to divorce. Patriarchal beliefs and institutions legitimate these inequalities and endorse dominant masculinity. Low-status, physically demanding labor along with caregiving fall mostly upon women and are avoided by men, whose reduced involvement with children lessens the modulating impact of hormonal changes on aggressive behavior discussed in Chapter 5.

To Graeber and Wengrow (2021) a change in women's freedom to escape an abusive household was key to the confluence of the three forms of domination beginning around 5,000 BP. Second-order kingdoms were modeled upon patriarchal households, with the public envisioned as family members subject to the king's paternal authority and responsibility. Kings provided a refuge and livelihood to vulnerable populations, including outcasts, women, orphans, and the disabled or ill. The work these individuals were given was considered honorable because it benefited the gods.

Third-order rulers set themselves up in the same courts and temples, or next door, and turned them into sites of exploitation. For instance, ambitious kings repurposed Sumerian temples such as the temple of Inanna (Ishtar) in Uruk into a sort of poorhouse for war captives and slaves. By removing the protectiveness and dignity of these sites, the change in temple life closed off a route to safety for women and thereby disempowered them within the household. Loss of the third freedom followed close behind the first two, casting women who dared to challenge the social order as dangerous subversives (Graeber and Wengrow 2021). Stone (1978) affirms that a comprehensive effort to suppress and erase female-centered religious leadership and traditions including worship of Inanna was essential to the expansion of a male-centered monotheistic politico-religious system in the ancient Near East and beyond.

Another piece of the puzzle involves the way states internalize violence, which Graeber and Wengrow (2021) explain through a comparison between 17th-century Iroquoian and European uses of community sanctioned violence. While women and children captured in war were incorporated into Iroquoian society, the treatment of men oscillated between extreme cruelty and complete assimilation depending on the decisions of women. Men who were brought in would be given a name that could even be that of the man whose death provoked the violence, whereas those sentenced to torture would suffer for days before dying of their injuries.

The public spectacle involved everyone as witnesses or deliverers of torments, even though the same people performing such sadistic acts brooked no form of it amongst themselves. Contemporaneously in France and elsewhere in Europe, the same sadism was applied to criminals and outcasts within society, reminding everyone that they could be its next victim. In the latter case, the threat of state violence limited or precluded the freedom to disobey, along with both physical and existential mobility, which, in turn, closed off the freedom to envisage and construct different ways of organizing society.

Other factors that contributed to the emergence of third-order polities include a shift in concepts of property ownership away from custodianship to full rights over the disposition of land and other possessions; the standardization of value through money or units of goods, which detaches economic relationships from the personalized exchanges in systems of reciprocity and redistribution; and, related to these two, the generation of poverty and debt due to the variability in individual fortunes that destines some people to fail to meet obligations of tax and tribute. To illustrate, both the Inca Empire and Andean village associations (*ayllu*) used knotted strings, *khipu*, to record units of value (Graeber and Wengrow 2021). In the first case, the knots kept track of labor obligations owed to the central administration and were never undone, whereas in the second they were unraveled upon repayment of a debt and all accounts between households were settled at the end of the year. The *ayllu* also redistributed farmland among households as they changed in size to ensure that none would be better off than others. In a similar manner, the *cargo* system of year-long civil/religious roles in Mesoamerica, which endured until the 1960s, counteracted the intrafamilial transmission of wealth. The system appears to have originated in a method devised by Spanish priests for delegating tasks to indigenous men. High-level *cargos* carried heavy financial burdens for masses, feasts, and saints-day festivals, which prevented men from working only for their own self-interest. If the system failed to redistribute a wealthy family's capital, the successive generation made more substantial contributions to religious observances (see Cancian 1974).

Graeber and Wengrow (2021) explain that early egalitarian settlements created a sense of belonging through the concept of a people as opposed to family or clan, which was key to the creation of civic institutions. For monarchies, the unifying narrative focused on divine or hereditary rulers. States, in contrast, elevate the idea of the nation. Demographic practices such as elections or lotteries to fill political positions mediate the sentiment of belonging and shift charismatic competitive politics from nobles and kings to the representatives of a similarly sidelined people.

Although the sources and consequences of the lost freedoms may be complex and obscure, people in a position to benefit from inequality readily understand how to manipulate its gears. Free-riders and ambitious individuals are kept in check in egalitarian societies, whereas self-serving and self-aggrandizing behavior pays off in unequal ones. Individuals and institutions protect their advantages by entrenching stratification and reinforcing beliefs about inherent differences that justify unequal access to wealth and power. Nationalistic ideologies create incompatibility with other societies, rationalizing international exploitation and warfare. The threat of natural disasters, external aggression, and other challenges allows leaders and elites to further chip away at the three freedoms and solidify their own advantaged position. It is in their interest to reduce the social trust that allows egalitarian societies to manage resources without authoritarian leaders, resolve disputes without formal courts, and redistribute their way through hard times.

## Ideologies of inequality

This chapter has shown, firstly, that humans are not biologically programmed for self-interested greed or unequal, competitive, violent societies, nor did agriculture and permanent settlement set humankind on a single relentless path leading to contemporary states. Evidence includes the biologically rewarding cooperative behavior and interest in strangers exhibited by preverbal children as well as nonhuman primates, which is accompanied by the expectation of fairness. Secondly, the egalitarianism of the only kind of human society that existed until the Holocene is culturally and socially built and maintained, although also favored by the demographic and ecological conditions associated with nomadic foraging. Open access to resources together with interconnectedness between groups gives foragers no need to fear strangers, narrows the range of reasons for conflict, and encourages avoidance above confrontation. Thirdly, the variability in the organization of human societies over the last 12,000 years shows that social and gender inequalities are human constructions. Many enormous early settlements got by for hundreds or thousands of years without monarchs, palaces and temples, or centralized administrations; others tried hierarchical systems and then rejected them. Urban settlements were not a clear break but rather were continuous with earlier, periodic or seasonal agglomerations, and they preceded full commitment to agriculture. Agriculture could be associated either with egalitarian or hierarchical systems, and technologies arose before they were put to practical use. In short, complexification depended on context and historical period, could be temporary, and was neither inevitable nor irreversible, allowing people to experience different sociopolitical systems, reevaluate, and revise.

Genetic essentialism is one of the many ideological techniques devised by human societies to rationalize inequality. Where some cultures attribute divine favor or divinity itself to kings and religious leaders while relegating commoners to lower status and worse living conditions, others construe such differences in terms of the separate biological properties of race and sex categories. In democratic societies, inequalities in personal agency, power, wealth, living conditions, and safety contradict avowed values of equality, solidarity, and liberty. The ideal of individual self-determination clashes with structural inequalities, engineered brakes on social mobility, and systems of domination and control from which there is no escape. Genetic essentialism resolves these contradictions by rendering disparities natural and unavoidable (see Figure 7.2). By attributing unequal opportunities and outcomes to the diversity in human biological endowments, genetic essentialism obscures the role of social arrangements.

Genetic essentialism affects intergroup interactions, willingness to address health and socioeconomic disparities, perceptions of human variation, the ability to learn about biology, educational and professional pathways, scientific research, and medical care. The current political climate in the US and elsewhere welcomes essentialist reasoning as a justification for undoing previous efforts to address historical and ongoing racism and sexism. Conservative political discourse insists, against scientific consensus, that sex is identical to gender and that, while racism does not exist, race does. These ideas drive efforts to repudiate transgender rights, dismiss leaders in government and the military based on skin color and gender, and dismantle programs and research aimed at documenting and ameliorating inequalities. Words and images that recognize the achievements of non-White and women citizens are erased from the public domain, while those which glorify White men are reinstated or introduced anew. The effect on

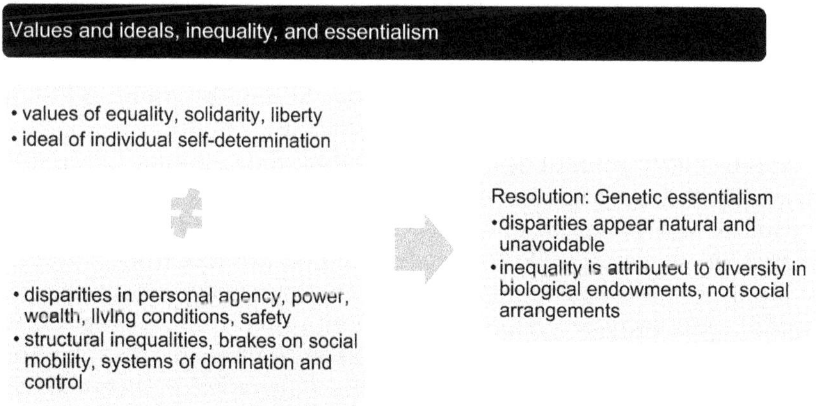


*Figure 7.2* Genetic essentialism resolves the contradiction between inequality and democratic ideals.

language and the visual landscape is only multiplied by information technologies that draw upon electronic resources.[10] At the same time, the genetic, biomarker, and personal behavior databases discussed in Chapter 3 reinforce genetic essentialism by design through categorization of individuals according to race and sex, with potentially harmful economic, educational, criminal justice, and health impacts for disadvantaged groups. Algorithms contribute to these same wide-ranging impacts while perpetuating discrimination against women and non-White individuals by directing internet traffic toward male and White contributors. The more disparities appear fixed and reflective of essential differences, the more they support discriminatory attitudes and practices, and the more they fuel online and in-person threats and hate speech.

What could break the chain connecting essentialism and inequality? The material presented in these pages points to the importance of anti-determinist, anti-essentialist science education and communication. The popularity of genetic determinism in the wider society needs to be countered through information on the complexities of phenotypic causation, which could temper the current predilection for describing the behavior of groups or even companies as being "in their DNA." Textbooks should clarify the meaning of heritability, avoid using racialized examples to illustrate genetic inheritance, explain human migrations and genetic variation within versus between populations, and address racism in science and society. Discussions related to sex should distinguish between sex and gender, explain the spectrum of sexes and genders, and describe the variability in gender roles and behavior within and between societies. Sections on human evolution should counter stereotypes about race, gender, and inequality through discussions of food procurement and social organization in prehistory, the evolution of prosocial traits, the variability in sociopolitical arrangements across time and space, the lack of population differences in mental traits, and the wrongheadedness of evolutionary progressions. Together these methods of presenting human variation would provide a more accurate understanding of both essentialism and inequality as human creations.

At the level of political action, governments could promote or dampen inequality through caps on executive pay, changes to tax codes, and expanded state provision of infrastructure, utilities, and social services. They could impose or undo legal codes and practices that deny or restrict access by nondominant groups to jobs, housing, credit, education, and medical care. Increased financial and social support could be given to informal organizations that already maintain egalitarian principles and structures which emphasize reciprocity and collaborative decision-making, such as recreational sports teams, charitable organizations, and neighborhood associations that manage administrative functions, organize informal policing, and provide care to vulnerable populations.

Societies that are more successful at equalizing social and economic conditions across the board are better able to support social connectedness and prevent the biological stress response activated by discrimination. Their populations are healthier. There is less gender, interpersonal, and intergroup violence, as well as less indirect structural violence through systems that subject disadvantaged groups to higher rates of poverty, food insecurity, disease, workplace injuries, and crime.

The simultaneous loss of the three freedoms described above leaves most of the world trapped in states and unable to create an alternative sociopolitical system. This chapter has shown that stratification and its impacts are rooted in knowable social, ecological, ideological, and demographic forces, rather than an imagined biological/evolutionary scenario. Confronting genetic essentialism could be a way to open a conceptual door to something new.

## Notes

1 In all three experiments, three-year-olds differentiated between the cooperation condition and the others. In two of the experiments (1 and 3), the two-year-olds did the same, but they did not differentiate in experiment 2.
2 In the first set of tests, on average 1.29 hours had passed since the last feeding versus 2.21 hours in the second experiment.
3 Non-reproductive socio-sexual behavior occurred in a portion of cases, involving genital-to-genital rubbing, but was not the motivator of the sharing behavior.
4 $n$ = 17,303; 355 cities.
5 The average score for servitude was 1.1 for mobile bands and 2.2 for settled foragers, where 1 signifies absence of servitude, 2 = incipient servitude, and 3 = hereditary slavery.
6 For a refutation of the argument that intergroup violence is unrelated to sociopolitical complexity based on prehistoric burials in California, see Fry et al. (2020).
7 Regarding exploitation of aquatic environments in this era, see Romàn et al. (2022).
8 Regarding pre-Holocene techniques for increasing the availability of desired foods, see Summerhayes et al. (2010); Thompson et al. (2021). Regarding plant and animal domestication, see Diamond (2002).
9 The urban settlements of Mesoamerica did not use metal tools, horses and similar animals, or wheeled transportation until after the arrival of Europeans.
10 To illustrate, commemorative coin designs for the 250th anniversary of the US, selected by a congressionally mandated committee in collaboration with government agencies, historians, and the public, include representations of women's suffrage, abolition (Frederick Douglass), and public-school integration (Ruby Bridges) (Felton 2025). These designs have been rejected in favor of images of US presidents and a pilgrim couple. The latter image shows a man dominant in size, gaze, and posture, one hand clasped over the woman's, the other grasping her shoulder. The woman wears a suffering expression and has neither hand visible.

## References


Alesina A, Giuliano P, Nunn N. 2013. On the origins of gender roles. *The Quarterly Journal of Economics* 128(2):469–530.

Baden AL, Wright PC, Louis Jr EE, Bradley BJ. 2013. Communal nesting, kinship, and maternal success in a social primate. *Behavioral Ecology and Sociobiology* 67(12):1939–1950.

Barragan RC, Brooks R, Meltzoff AN. 2020. Altruistic food sharing behavior by human infants after a hunger manipulation. *Scientific Reports* 10(1):1785.

Bird DW, Bird RB, Codding BF, Zeanah DW 2019. Variability in the organization and size of hunter-gatherer groups. *Journal of Human Evolution* 131:96–108.

Blessin M, Lehmann S, Kunzler AM, van Dick R, Lieb K. 2022. Resilience interventions conducted in western and eastern countries. *International Journal of Environmental Research and Public Health* 19(11):6913.

Brown SL, Nesse RM, Vinokur AD, Smith DM. 2003. Providing social support may be more beneficial than receiving it. *Psychological Science* 14(4):320–327.

Burkett JP, Andari E, Johnson ZV, Curry DC, de Waal FB, et al. 2016. Oxytocin-dependent consolation behavior in rodents. *Science* 351(6271):375–378.

Cancian F. 1974. New patterns of stratification in the Zinacantan cargo system. *Journal of Anthropological Research* 30(3):164–173.

Cashdan, Elizabeth A 1980. Egalitarianism among hunters and gatherers. *American Anthropologist* 82(1):116–120.

Chagnon NA. 1968. *The fierce people*. Holt, Rinehart & Winston.

Cohen MN. 1989. *Health and the rise of civilization*. Yale University Press.

Cohn A, Maréchal MA, Tannenbaum D, Zünd CL. 2019. Civic honesty around the globe. *Science* 365(6448):70–73.

Crevecoeur I, Dias-Meirinho MH, Zazzo A, Antoine D, Bon F. 2021. New insights on interpersonal violence in the Late Pleistocene based on the Nile valley cemetery of Jebel Sahaba. *Scientific Reports* 11(1):9991.

De Waal F, Lanting F. 1997. *Bonobo*. University of California Press.

Diamond J. 2002. Evolution, consequences and future of plant and animal domestication. *Nature* 418(6898):700–707.

Duffy PR, Wilkes F, Skorna H, Furholt M, Dickie C, et al. 2025. Five thousand years of inequality in the Carpathian Basin. *Science Advances* 11(32):eadu0323.

Dunbar RI. 2003. The social brain. *Annual Review of Anthropology* 32(1):163–181.

Felton E. 2025. US anniversary coins won't feature any Black Americans or notable women. 20 September. https://www.msn.com/en-us/society-culture-and-history/history/us-anniversary-coins-won-t-feature-any-black-americans-or-notable-women/ar-AA1SJ5uu

Friedl E. 1978. Society and sex roles. *Human Nature* 1(4):68–75.

Fry DP. 2006. *The human potential for peace*. Oxford University Press.

Fry DP, Keith CA, Söderberg P. 2020. Social complexity, inequality and war before farming. In Moreau P, editor. *Social inequality before farming?* McDonald Institute for Archeological Research. pp. 303–320.

Fry DP, Söderberg P. 2013. Lethal aggression in mobile forager bands and implications for the origins of war. *Science* 341(6143):270–273.

Gesquiere LR, Adjangba C, Young G, Brandon C, Parker S, et al. 2025. Energetic costs of social dominance in wild male baboons. *Proceedings of the Royal Society B: Biological Sciences* 292(2039).

Ghosh P. 2022. Neanderthal extinction not caused by brutal wipeout. *BBC News*. https://www.bbc.com/news/science-environment-60305218

Gilby IC, Brent LJ, Wroblewski EE, Rudicell RS, Hahn BH, et al. 2013. Fitness benefits of coalitionary aggression in male chimpanzees. *Behavioral Ecology and Sociobiology* 67(3):373–381.
Graeber D, Wengrow D. 2021. *The dawn of everything*. Picador.
Gretzinger J, Schmitt F, Mötsch A, Carlhoff S, Lamnidis TC, et al. 2024. Evidence for dynastic succession among early Celtic elites in Central Europe. *Nature Human Behaviour* 8(8):1467–1480.
Grossman D. 2014. *On killing*. Open Road Media.
Haas J, Piscitelli M. 2013. The prehistory of warfare. In Fry DP, editor. *War, peace, and human nature*. Oxford University Press. pp. 168–190.
Hamann K, Warneken F, Greenberg JR, Tomasello M. 2011. Collaboration encourages equal sharing in children but not in chimpanzees. *Nature* 476(7360):328–331.
Haraway DJ. 2013. *Primate visions*. Routledge.
Hill KR, Wood BM, Baggio J, Hurtado AM, Boyd RT. 2014. Hunter-gatherer inter-band interaction rates. *PloS one*. 9(7):e102806.
Hodgson S, Watts I, Fraser S, Roderick P, Dambha-Miller H. 2020. Loneliness, social isolation, cardiovascular disease and mortality. *Journal of the Royal Society of Medicine* 113(5):185–192.
Hoebel EA. 1954. *The law of primitive man*. Harvard University Press.
Horner V, Carter JD, Suchak M, De Waal FB. 2011. Spontaneous prosocial choice by chimpanzees. *Proceedings of the National Academy of Sciences* 108(33):13847–13851.
Hublin JJ, Sirakov N, Aldeias V, Bailey S, Bard E, et al. 2020. Initial Upper Palaeolithic *Homo sapiens* from Bacho Kiro Cave, *Bulgaria. Nature* 581(7808):299–302.
Huchard E, Kappeler PM, Smit N, Fichtel C, Lukas D. 2025. The evolution of male–female dominance relations in primate societies. *Proceedings of the National Academy of Sciences* 122(28):e2500405122.
Hung LW, Neuner S, Polepalli JS, Beier KT, Wright M, et al. 2017. Gating of social reward by oxytocin in the ventral tegmental area. *Science* 357(6358):1406–1411.
Koptekin D, Aydoğan A, Karamurat C, Altınışık NE, Vural KB, et al. 2025. Out-of-Anatolia. *Science* 388(6754):eadr3326.
Krasniqi S, Krasniqi B. 2019. Sport and peacebuilding in post-conflict societies. *Journal of Aggression, Conflict and Peace Research* 11(3):145–157.
Lacy S, Ocobock C. 2024. Woman the hunter. *American Anthropologist* 126(1):19–31.
Marmot MG, Shipley MJ, Hemingway H, Head J, Brunner EJ. 2008. Biological and behavioural explanations of social inequalities in coronary heart disease. *Diabetologia* 51(11):1980–1988.
Masclans Latorre A, Bickle P, Hamon C. 2021. Sexual inequalities in the Early Neolithic? *Journal of Archaeological Method and Theory* 28(1):232–273.
Mauss M. 1990 [1923–24]. *The gift: The form and reason for exchange in archaic societies*. Hall WD, translator. Routledge.
Meijer H. 2024. The origins of war. *Human Nature* 35(3):225–288.
Mitani JC, Watts DP, Amsler SJ. 2010. Lethal intergroup aggression leads to territorial expansion in wild chimpanzees. *Current Biology* 20(12):R507–508.
Mittnik A, Massy K, Knipper C, Wittenborn F, Friedrich R, et al. 2019. Kinship-based social inequality in Bronze Age Europe. *Science* 366(6466):731–734.
Montagu A, editor. 1978. *Learning non-aggression*. Oxford University Press.
Nolan PD. 2003. Toward an ecological-evolutionary theory of the incidence of warfare in preindustrial societies. *Sociological Theory* 21(1):18–30.

Rogers DS, Deshpande O, Feldman MW. 2011. The spread of inequality. *PLoS One* 6(9):e24683.
Román D, Martínez-Andreu M, Aguilella G, Fullola JM, Nadal J. 2022. Shellfish collectors on the seashore. *The Journal of Island and Coastal Archaeology* 17(1):43–64.
Sahlins M. 2013. *Stone age economics*. Routledge.
Schülke O, Bhagavatula J, Vigilant L, Ostner J. 2010. Social bonds enhance reproductive success in male macaques. *Current Biology* 20(24):2207–2210.
Sommerville JA, Schmidt MF, Yun JE, Burns M. 2013. The development of fairness expectations and prosocial behavior in the second year of life. *Infancy* 18(1):40–66.
Staes N, Stevens JM, Helsen P, Hillyer M, Korody M, Eens M. 2014. Oxytocin and vasopressin receptor gene variation as a proximate base for inter-and intraspecific behavioral differences in bonobos and chimpanzees. *PLoS One* 9(11):e113364.
Stone M. 1978. *When God was a woman*. Houghton Mifflin Harcourt.
Stoneking M, Arias L, Liu D, Oliveira S, Pugach I, et al. 2023. Genomic perspectives on human dispersals during the Holocene. *Proceedings of the National Academy of Sciences* 120(4):e2209475119.
Summerhayes GR, Leavesley M, Fairbairn A, Mandui H, Field J, et al. 2010. Human adaptation and plant use in highland New Guinea 49,000 to 44,000 years ago. *Science* 330(6000):78–81.
Sun F, Wu YE, Hong W. 2025. A neural basis for prosocial behavior toward unresponsive individuals. *Science* 387(6736):eadq2679.
Tan J, Hare B. 2013. Bonobos share with strangers. *PLoS One* 8(1):e51922.
Thompson JC, Wright DK, Ivory SJ, Choi JH, Nightingale S, et al. 2021. Early human impacts and ecosystem reorganization in southern-central Africa. *Science Advances* 7(19):eabf9776.
Wang Y, Ge J, Zhang H, Wang H, Xie X. 2020. Altruistic behaviors relieve physical pain. *Proceedings of the National Academy of Sciences* 117(2):950–958.
Whiting BB. 1975. Aloofness and intimacy of husbands and wives. *Ethos* 3(2):183–207.
Widlock T. 2002. The long walk IV-Hunter-gatherers and anthropology. *Nomadic Peoples* 6(1):1–19.
Yüncü E, Doğu AK, Kaptan D, Kılıç MS, Mazzucato C, et al. 2025. Female lineages and changing kinship patterns in Neolithic Çatalhöyük. *Science* 388(6754):eadr2915.
Zeanah DW, Elston RG, Codding BF. 2026. Gender convergent labor and technological change at the Pleistocene to Holocene Transition in the Great Basin. *Journal of Anthropological Archaeology* 81:101746.

# Index

Pages in *italics* refer to figures, pages in **bold** refer to tables, and pages followed by n refer to notes

For Product Safety Concerns and Information please contact our EU
representative GPSR@taylorandfrancis.com
Taylor & Francis Verlag GmbH, Kaufingerstraße 24, 80331 München, Germany

www.ingramcontent.com/pod-product-compliance
Ingram Content Group UK Ltd.
Pitfield, Milton Keynes, MK11 3LW, UK
UKHW021306260926
475702UK00048B/740
*9781041073161*